DEBATING DISASTER RISK

DEBATING DISASTER RISK

ETHICAL DILEMMAS IN THE ERA OF CLIMATE CHANGE

EDITED BY
GONZALO LIZARRALDE,
LISA BORNSTEIN, AND
TAPAN DHAR

Columbia University Press *New York*

Columbia University Press
Publishers Since 1893
New York   Chichester, West Sussex

Library of Congress Cataloging-in-Publication Data
Names: Lizarralde, Gonzalo, editor. | Bornstein, Lisa, editor. |
Dhar, Tapan, editor.
Title: Debating disaster risk : ethical dilemmas in the era of climate change /
edited by Gonzalo Lizarralde, Lisa Bornstein, and Tapan Dhar.
Description: New York : Columbia University Press, 2025. |
Includes bibliographical references and index.
Identifiers: LCCN 2025008640 (print) | LCCN 2025008641 (ebook) |
ISBN 9780231209663 (hardback) | ISBN 9780231209670 (trade paperback) |
ISBN 9780231558051 (ebook)
Subjects: LCSH: Emergency management. | Climatic changes.
Classification: LCC HD49 .D427 2025 (print) | LCC HD49 (ebook)

Cover design: Noah Arlow
Cover image: Shutterstock

GPSR Authorized Representative: Easy Access System Europe,
Mustamäe tee 50, 10621 Tallinn, Estonia, gpsr.requests@easproject.com

Dedicated to Graham Saunders (1961–2017), with thanks for his lifesaving contributions to this area of study

CONTENTS

PREFACE

Over the past few years, our team at the Canadian Disaster Resilience and Sustainable Reconstruction Research Alliance (*Œuvre Durable* in French) has realized that the disaster research field is full of "good ideas" that often fail to produce positive and long-term change. Why is it so difficult to create positive change, even when we act with good intentions? Why is it so hard to reduce vulnerabilities and achieve security and protection? Why are we failing to protect humans from natural hazards and ecosystems from human threats?

Puzzled by these questions, we invited twenty-two urban consultants, academics, practitioners, humanitarian agents, and politicians across the globe to rethink the simplistic adoption of existing frameworks and engage in a more thorough examination of ethics in the disaster field. We focused on interconnections among risk, disasters, and the built environment, including issues related to city planning, urban design, infrastructure development, and housing. The experts participated in eleven online debates, arguing different viewpoints and examining common controversies in urban development, risk reduction, and post-disaster reconstruction. In each debate, we animated a conversation by engaging two internationally known experts, a moderator, hundreds of people who voted on the debate motion, and dozens of students and scholars who commented on the subject.

The debates became a huge success, forming a platform for engaged knowledge sharing among professionals, academics, practitioners, and students worldwide. The platform attracted more than 34,000 visits and 14,000 visitors, thousands of votes, and hundreds of comments. In *Debating Disaster Risk: Ethical Dilemmas in the Era of Climate Change*, we compile and analyze the results of this work.

Building on multidisciplinary views on disaster risk reduction and response, we explore common controversies and conflicting ideas often adopted by academics, authorities, and international agencies. We forward a comprehensive view of risk and disasters, exploring their political dimensions and connections to social, cultural, and economic systems, but we focus particularly on the controversial interconnections among society, nature, and the built environment. Issues related to housing, environmental protection, urban development, informal forms of construction, infrastructure recovery, and urban systems emerge almost naturally in all debates. We put less emphasis, however, on other issues such as health, energy production, reductions in carbon emissions, insurance, and industrial development. These are surely crucial aspects of vulnerability reduction, but we consider them as variables interacting with society, nature, and the built environment, which emerge in our book as key domains within which almost all other human activities take place.

We understand radical change and destruction as opportunities to learn about culture, power, and institutions. We try to unpack the complexity of decisions required in a world where social injustices persist and environmental damage is rampant. We invite you to examine the advantages and pitfalls of popular concepts and frameworks and to challenge conventional thoughts and concepts. Our arguments are supported by scientific work and framed by the political and ethical stances of our editorial team. We invite readers to incorporate their own ethical judgment and political perspective in their understanding of the most pressing current environmental and social problems.

ACKNOWLEDGMENTS

This book is inappropriately signed by only three editors. In reality, the method, arguments, and results presented in *Debating Disaster Risk: Ethical Dilemmas in the Era of Climate Change* have been developed by many people over the past eight years. We acknowledge the significant contribution of all members of the *Œuvre Durable* research team who initiated and supported oddebates.com, the online platform of live debates that became the foundation of this book. Special thanks are given to Kevin Gould, Gabriel Fauveaud, Christopher Bryant, Nalini Mohabir, David Wachsmuth, and Danielle Labbé, who significantly contributed to the design and animation of the debates.

This publication and the online debates were funded by Fonds de recherche du Quebec–Sociéte et culture (FRQSC), and the Social Sciences and Humanities Research Council of Canada. We thank both foundations for their valuable financial support.

We also extend our sincere gratitude to the authors, students, and visitors who participated in the online debates. Our appreciation extends to professors, researchers, collaborators, and practitioners for contributing to the book in different ways, notably:

- The organizers of the online debate series: Mahmood Fayazi (2014–2018), Faten Kikano (2019), and Mauro Cossu (2020–2023), who formed the backbone of the online platform and collaborated with all the authors and hundreds of participants.
- The members and contributors of the i-Rec network (Information and Research for Reconstruction) for their intellectual support and collaboration during the online debates.
- Students on the Université de Montréal research team, in particular Anne-Marie Petter, Arturo Valladares, Benjamin Herazo, Faten Kikano, Gabriela Gonzales Faria, Lisa Hasan, Mauro Cossu, Steffen Lajoie, and many others who inspired us with their innovative ideas, timely suggestions, and support.
- The invited researchers and professors in our team, including Holmes Julián Páez Martínez, Tamiyo Kondo, and Ernesto Aragón.
- Students on the McGill University research team, including Roger Ishimwe, Sayana Sherif, Simon Laflamme, and Waqas Ahmed Raza.
- Kate Stern, for revising and commenting on the initial book proposal, and Victoria Addona and Jordan Larson, who provided us with detailed edits and suggestions at the final stage of manuscript preparation.
- Residents, local leaders, and agencies who participated in our case studies and shared their stories and experiences.

We also thank Robert Lecker, our literary agent, for helping us at different stages of the publication process, and Lowell Frye and Miranda Martin, the editors at Columbia University Press, for providing us with practical guidelines for preparing the manuscript.

Lastly, we are grateful to all members of the ADAPTO and SUSTENTO projects, funded by Canada's International Development Research Centre (IDRC).

INTRODUCTION

The Problem of Explaining Problems

TAPAN DHAR, LISA BORNSTEIN,
AND GONZALO LIZARRALDE

WE WOULD RATHER BE KILLED BY STORMS THAN SLOWLY STARVE

We visited Jelepolli, a fishing village on the bank of the Baleshwar River in Sharankhola, Bangladesh, on a cloudy and windy afternoon. The vast water, hazy skies, and colorful homes created a stunning landscape (see figures 0.1 and 0.2). We wanted to know about the villagers' lives, their struggles, their stories, and their ideas for the future. Amena Begum shared one of those stories:

> In the afternoon of a windy day, my husband was in Dhaka and I was sitting on the veranda with the children, like on many other rainy days. A neighbor told us that the water level of the Baleshwar was increasing and flowing toward us. I thought the water surge would not be that bad. It surely would be like last year's, which damaged the house floor and walls. But then, the whole house was shaking, neighbors were shouting, and the Imam started praying. I didn't wait any longer, held my daughter's hand, and tied my son to my waist. Most of us tried to find shelter in the Mosque. But the building was already destroyed—the roof

> was gone. A high wave of water took us. I tried to hold my son but could not. I do not recall what happened to my daughter and nephew. I tried to swim but lost consciousness.

Sharankhola is one of the lowest-lying areas in southwest Bangladesh, constantly exposed to cyclones and water surges, such as Cyclone Bulbul in 2019 and Cyclone Amphan in 2020. Referring to the 2009 Cyclone Aila, Amena continued: "After many hours, perhaps the following morning, they found me with many others on the banks of the Baleshwar. I was still unconscious, but Thank God, alive. Later, I found my nephew, who managed to hold on to a tree and saved himself." Three days later, Amena's daughter was found dead a kilometer away from the village.

Sadly, Amena's experience is not unique. Her neighbor Ruhul Amin suffered firsthand the destruction caused by Cyclone Sidr in 2007 and Cyclone Amphan in 2020. The first cyclone killed his two sons; the second caused a huge financial burden. Fishermen like Ruhul go to the sea in groups of six and stay away from home for about seven days, renting fishing boats because they cannot afford to buy their own. Depending on the weather, they fish once or twice a month. The day Cyclone Amphan hit, Ruhul and his group were fishing. As they prepared to return to Jelepolli, they noticed changes in the sea and the afternoon sky. They were near the estuary in the Sundarban area, out of the range of cell phone coverage. They rowed to a narrow canal and anchored their six boats. Leaving behind a complete load of fish, they took some dry food and entered the deep forest in search of protection. The next morning, after the cyclone hit, Ruhul found only two boats, both broken. Two partners were injured and could not work for several weeks. The group returned to the village with no fish, few possibilities for work in the following

FIGURE 0.1 View of the Baleshwar River.

months, and a pressing problem: they had to pay for the broken and missing boats.

Almost all residents of southwest Bangladesh have experienced cyclones and storms. Like Ruhul, most are familiar with their early signs and know what to do during and after such events. Repairing houses and infrastructure is a regular activity for villagers in the region—Ruhul has rebuilt his home five times in fifteen years. But to do so, most villagers borrow money from local moneylenders at high interest rates. As villagers like Ruhul struggle to secure a stable income, their debts are becoming a serious burden.

Of course, many in Jelepolli believe that cyclones are cruel, albeit temporary. Economic impacts endure longer than the natural events, while poverty has become permanent. Sometimes

FIGURE 0.2 A traditional house in Sharankhola, made of mud and wood and featuring a raised floor to cope with regular water surges.

villagers find that debt, lack of food, and loss of income are more dangerous than cyclones. They dislike that few aid agencies and government representatives consider programs to improve their livelihoods after disasters hit. Instead, aid initiatives built cyclone shelters and hundreds of kilometers of embankment infrastructure along major rivers.[1] To be fair, efforts have been made to adopt new technologies for early warning systems and risk awareness programs, though they remain insufficient—for instance, less than one-fifth of the population in the region has access to cyclone centers.[2]

When we visited Jelepolli, the government was building a sixty-two-kilometer embankment along the Baleshwar River. Many residents feared that, instead of mitigating problems, the new infrastructure would cause environmental problems and increase risks.[3] After the construction of similar embankments in the territory, rivers have run dry, floods have appeared in other areas, and waterlogging of soils has become more common. Locals know that this type of infrastructure reduces the benefits of monsoon rains and regular crop inundations. Some of these projects affect the ecosystem required for the unique deltaic agriculture of the region, crippling the local economy. As a result, food insecurity has become a pressing problem, particularly during the monsoon season.

Villagers are afraid of floods but face other challenges on a daily basis. In the aftermath of the cyclone, Amena, like other villagers, continues to fish and cultivate orchards and food gardens on nearby embankments. Securing a consistent income has become almost impossible and yet Ruhul, Amena, and other villagers are resisting relocation. They know that their agriculture and fishing skills are not useful in other places, particularly in Dhaka and other large cities. When asked about the possibility of moving to a safer area, one resident told us, "We would rather be killed by storms than slowly starve later on."

In Jelepolli, most disaster risk reduction activities spearheaded by government and aid agencies have focused on protecting lives and assets, mostly by building shelters and infrastructure. New construction projects attract donors, some of them motivated to access what NGOs have called a laboratory for "resilient development."[4] In many cases, the root causes of vulnerability have been insufficiently addressed. Today, Ruhul and Amena have better chances of obtaining emergency shelter than proper education for their children, food security, or a solution to their lack of income. Such imbalanced responses partially derive from the ways government, NGOs, and international organizations frame the problem, confront conflicting interests, and implement change.

Many organizations working in Bangladesh rely on narratives of resilience, sustainability, climate adaptation, green development, technological innovation, and community participation in their efforts to attract donors and influence policy. But what are the impacts of the adoption of such concepts? What do such ideas, often imported from elsewhere, reveal about lived reality in Jelepolli and similar villages along the Baleshwar River? Do these notions help us understand the daily struggles and emotions faced by Ruhul and Amena, and their attachment to place and cultural traditions?

This book aims to provide answers to these and similar questions and demands a reassessment of ethical stances in responses to disasters and risks. We hope that by revealing the complexity of problems linked to disaster risk reduction, adaptation, resilience, and aid, decision-makers will be equipped to provide better solutions to people like Ruhul and Amena.

UNDERSTANDING COMPLEXITY

In this book, we contend that dealing with disasters, climate change, and destruction is a political process.[5] Responses taken for risk reduction or climate action can enhance the mobility or permanence of populations, provide relief, or lead to increased suffering. The outcomes of such processes create winners and losers, producing sometimes radical changes or other times continuity in livelihoods and urban systems. As illustrated by the case of Bangladesh, responses to disasters and climate change are a matter of life or death for millions of people worldwide.

Adopting a good idea, one often linked to resilience, adaptation, sustainability, participation, technical innovation, or similar concepts,[6] is thus an attractive option for the many government representatives, consultants, humanitarian agents, and scholars who deal with threats and destruction. But often such ideas are overly simplistic, inadequate to the complexity of the challenges.

For example, commonly circulated ideas are that cities must become more intelligent and resilient; disaster victims must be more engaged and participatory; emergency shelters must be prefabricated and assembled more quickly; the wealthy should donate more; and buildings and infrastructure must be greener. In other words, disaster risk reduction and aid are marked by an interest in adopting normative approaches to "fix" current problems. But most of these ideas, though they seem promising on the surface, hinder a more thorough analysis of cultural factors, disregard significant difficulties linked to implementation, perpetuate secondary effects, and mask some decision-makers' objectives and motivations.[7]

More worryingly, these ideas are typically implemented without proper adaptation to local contexts. They are insufficiently challenged and, as a consequence, proliferate in vague and general ways; they become buzzwords that fail to explain the complexity of current problems. Scholars have found, for instance, that Build Back Better has become a common but empty mantra used by governments and NGOs that does not describe what needs to be done in reconstruction projects, by whom, and how.[8] The concept of smart cities has come to indicate how to avoid risks and rebuild efficiently after destruction, but it rarely addresses the ethical, social, cultural, and political dynamics associated with the implementation of technological change in urban settings. In an era dominated by artificial intelligence solutions, governments and international organizations are investing millions of dollars in state-of-the-art technology and systems for risk reduction,[9] even if their efficiency remains to be demonstrated. The concept of community is called upon to refer to different entities, often in a manner that oversimplifies social affiliations, origins, and the complexity of relational dynamics within and across different groups.[10] Resilience, sustainability, and adaptation are used so widely, in so many ways and contexts, that they may not mean anything significant. In this book, we argue that in adopting these ideas without questioning their real value, we fail to understand the underlying rationales of decision-makers, how explanations of current problems create blind spots, and how the implementation of projects is often difficult and produces a host of secondary effects.

We contend that the way we frame current problems and the struggles lived by Ruhul, Amena, and millions of people worldwide is key to finding more appropriate solutions. As we shall see, setting the problem is, in itself, a problem.

FRAMING CURRENT AND FUTURE PROBLEMS AND THEIR SOLUTIONS

Today, disasters are four to five times more frequent than in the 1970s and cause seven times the damage.[11] Whereas some of the hazards that affect Ruhul and Amena are natural, the daily struggles and disasters that affect their village are not.[12] Disasters and climate change have disproportionate impacts on those who are poor, marginalized, and excluded, whether living in industrialized nations or the lowest income nations of the Global South. Black and indigenous communities, for instance, are more likely to suffer from disasters in both the Global North and South. Women, elderly people, and low-income households find it harder to recover after disasters almost everywhere.

Risk exposure and vulnerabilities are not distributed equally within and among territories, countries, and social groups. High-income nations, for instance, suffer from severe economic damages but count fewer deaths and injuries caused by disasters than nations in the Global South. It is estimated that as much as 91 percent of deaths linked to weather, climate, and water-related disasters between 1970 and 2019 occurred in developing countries.[13] About three-quarters of global disaster victims live in low- to medium-income countries.[14]

Whereas scientists have challenged the "natural" character of disasters for almost four decades,[15] authorities and analysts continue to blame disasters on the "unexpected" nature of hazards, geophysical determinants, or meteorological events.[16] For most contemporary scholars, the severity of disaster impacts on a particular society reflects social and environmental injustices and the sensitivity of human, economic, and political systems.[17] They argue that positivist and deterministic explanations of disasters, which typically overlook the sociopolitical construction of vulnerability, mask the actual causes of risk creation and its liability.[18]

Climate change and the unprecedented damage to ecosystems are challenging how we explain risks and the "natural" character of hazards.[19] Rapid population growth coupled with uncontrolled urbanization is responsible for the creation of new hazards such as heat islands, urban noise, and light, water, and air pollution.[20] We know that industrialization, deforestation, large-scale mechanized agriculture, air travel, car use, and the unprecedented rise in fossil fuel consumption contribute to carbon emissions and thus precipitate the conditions that have led to global warming.[21] Depletion of resources and human-induced changes in ecosystems are reducing our capacity to respond to hazards and creating new threats. But the causes and effects of global warming differ geographically. Whereas most atmospheric pollution occurs in the Global North, the effects of climate change are disproportionately felt by people in the Global South. Climate change affects not only human systems but also the fauna and flora. Its effect on nonhuman species remains understudied by disaster scholars, particularly those coming from the social sciences and humanities. Today, humans are increasingly at risk of being affected by natural hazards, and natural ecosystems are unprecedentedly at risk of being destroyed by human activity.

As human and urban systems become more complex, and their reach truly planetary, their impacts on natural systems, water, and ecosystems are intensifying.[22] The effects of disaster events are felt by more people, companies, and institutions, sometimes across countries and territories. But as infrastructure, mobility systems, telecommunications, and economic structures grow and interconnect, cause-effect explanations of disaster events are harder to identify. Understanding risks and responses to them is today a treacherous domain of study.

Scholars continue to claim that human capacities, aspirations, needs, experience, and preferences are overlooked in policy and action before and after disasters.[23] They find that stakeholders engaged in emergency aid do not always anticipate the long-term consequences of immediate responses. Emergency shelters tend to become permanent, with those temporarily built on the outskirts of cities contributing to urban sprawl and loss of agricultural or open lands.[24]

Research also shows that external aid also often disregards local perspectives, including, for example, sidelining indigenous concepts of nature and territory.[25] Grassroots and informal solutions frequently remain ignored in disaster risk reduction and climate policy.[26] Recent scholarship argues that disaster policy and vulnerability-reduction plans rarely deploy non-anthropocentric approaches to environmental problems.[27]

Disaster risk reduction and response demand collaborative actions at different scales and across various sectors. For decades, international agencies have promoted dozens of guidelines and pathways to this end, such as the 2005 Hyogo Framework for Action and the 2015 Sendai Framework for Disaster Risk Reduction.[28] However, Ben Wisner, an academic researcher at the Institute for Risk and Disaster Reduction, University College London, found that these frameworks remain "strikingly oblivious to root causes of disaster."[29]

Scholars have long acknowledged the difficulty of studying the conditions and root causes of vulnerabilities in disaster work.[30] A possible reason is the way such frameworks are dominated by "Western epistemologies."[31] Critics observe that research by authors outside of areas affected by disasters dominates disaster studies; for example, researchers from OECD[32] countries have contributed 80 percent of disaster studies over the past four decades.[33] Local researchers produced less than 10 percent of studies on recent disasters that occurred in developing countries such as Haiti, the Philippines, and Nepal. What are the consequences of outside research dominating most studies in disaster-affected areas? How does this unequal distribution of scientific work modify the ways we explain risks, disasters, and climate change effects?[34] As foreign researchers, our understanding of local narratives and explanations of local struggles remains limited; one response is to recognize that we must move from a focus on "our" science and pay more attention to *their* story.[35] We require a more comprehensive understanding of socioeconomic and political systems, as well as the actions of citizens, policymakers, and humanitarian agents, specific to case studies.

In this book, we invite scholars and practitioners to discuss further the consequences of adopting existing frameworks while examining the moral value of action in the field of disaster risk reduction. We bring together a multiplicity of voices and different perspectives to understand the complexity of dealing with risks, climate change, and disasters.

CREATING PERTINENT UNDERSTANDINGS THROUGH THE CONTRIBUTIONS OF MULTIPLE VOICES

This book aims to highlight the difficulties in making choices related to vulnerability reduction and post-disaster reconstruction. To do so, we bring together the arguments of twenty-two experts in disasters, reconstruction, and climate change. Each chapter is the outcome of a live ten-day online debate organized by our research team between 2017 and 2022. Each debate explores a controversy in disaster management and risk reduction, as articulated around a specific question.

In each debate, a moderator introduced a controversy, animated the discussion, and summarized the main conclusions of the activity. For each debate, we invited two panelists, recognized experts in the field, to present their most convincing arguments and counterarguments, for or against the topic, in three stages (introduction, rebuttal, and conclusions). They were asked to start with a strong stance; they then had the opportunity to respond to their opponent's views and nuance their positions.

In parallel, we invited the public to respond to the topic questions as well as the comments raised by the panelists. They could vote on the debate question at four different stages: before panelists' opening remarks, after their opening remarks, before their rebuttal remarks, and after their closing remarks. The debates adopted a dialectic method, co-constructing knowledge and ideas through several exchanges among the moderator, panelists, voters, and other participants. This approach is built on a rigorous examination of pros and cons, arguments and counterarguments, about risk, disasters, and climate change issues.

To select the panelists, we researched the identified topics, reviewed practical and scientific contributions of possible participants, and issued invitations. We also tried to bring together specialists from different disciplines, diverse backgrounds, and different locations. To reach our audience, we used academic, professional, and social networks including popular listservs such as i-Rec and Radix, and social media platforms such as LinkedIn, Instagram, and Facebook. Thousands of people joined the exercise, with the platform receiving about 34,000 visits from 14,000 visitors. Hundreds of online visitors commented and interacted with the moderator, panelists, students, and scholars. Most of the participants were professionals, academics, practitioners, and students from different parts of the world; almost all countries were represented. We obtained participants' consent to include excerpts of some comments in the book. We prioritized comments that had more influence in the debate and ensuing discussion. We hope that readers will sense participants' diversity, different voices, and unique observations stemming from their roles—whether as academics, practitioners, residents, or a combination thereof—and geographic locations. Each chapter includes a diagrammatic summary of what we know of participants' professional and geographic affiliations. We did not request and therefore do not present information on racial, gender, or other identity characteristics of our participants and panelists.

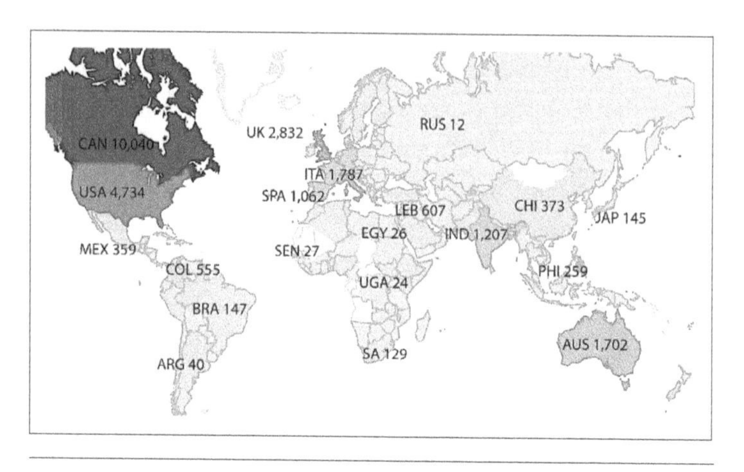

FIGURE 0.3 Number of visits per country during the online debates. We reached people from all continents and almost every country.

In each chapter, you will hear the voices of two world-class scholars or practitioners as they attempt to persuade you with their best arguments. You will also encounter the arguments of the debate moderator, who introduces and closes each debate, and several comments raised by our online audience. In some cases, we used multiple languages in our opening statements to reach a diverse range of audiences and to include participants' voices in their own languages. Finally, since participants were encouraged to vote on their positions, you will find the summarized results of their votes.

Given that the objective of this book is to bring together different perspectives, we have tried to maintain the original voice of panelists and participants, though light edits were required to ensure consistency in language and terminology. As much as possible, we have tried to respect the way panelists and participants used terms and concepts, which means that they have not been standardized as in a book or article written by a single set of authors. You will read, for instance, about issues in "poor countries," "developing countries," "low-income nations," and "countries of the Global South." We recognize that these terms often overlap, but they also translate different perceptions about lived reality. In many cases, such terms also exemplify how concepts and ideas have evolved (social scientists in 2024, for example, rarely refer to "poor countries" and instead use the notion of a "Global South").

The fact that we bring together different voices also means that concepts and ideas cannot be employed interchangeably. Terms such as "humanitarian aid," "emergency response," and "immediate disaster relief" are sometimes used to describe the same things, but they also carry nuances. "Humanitarian aid" tends to refer to a form of charity that is not always implied in the "immediate disaster relief" activities of local members of the civil defense. Similarly, "urban system" is not a synonym of "city infrastructure," but both terms might be used by debate participants to refer to traffic roads and railways. Debate participants from different countries also advance different notions of "international development," "economic growth," and "progress." We invite readers to appreciate the diversity of voices presented in the following debates and to regard the varied use of language as a manifestation of different worldviews. Whereas these differences might be interpreted as a lack of coherence in a standard academic book, we believe that here they contribute to the learning experience.

LANGUAGE, TOUGH CHOICES, AND WHAT OUGHT TO BE

Following this introduction, the book is structured in three parts, each responding to a key question: (1) What are the main tensions and problems in our interpretations of risk, disasters, and climate change? (2) Why and how do these explanations of risk affect our decisions in the face of disasters and climate change? (3) How should we address the most pressing problems raised by these issues?

The first section (part I) deals with the problem of explaining risk, disasters, and the responses to them. We include here three debates that explore language, meanings, and frameworks. How we define problems, and the language employed to explain them, matters. International frameworks are used to formulate disaster and climate policy, funding is allocated to problems as defined by stakeholders, and meanings determine not only how we see reality but also how we want to change it. The debates explore the ways in which cities are vulnerable; the influence of ideas from the Global North in disaster-related research, policy, and practices; and the usefulness of the notion of resilience. Each chapter constitutes an opportunity to question, and to challenge, the way we see problems and their solutions.

The second section, on tough choices (part II), explores tensions that emerge during responses to disaster- and conflict-induced displacement and the challenges of responding well. Over the past three decades, the frequency and intensity of disasters have substantially increased. Disasters kill fewer people today than decades ago, but their destructive impact on ecosystems, housing, and infrastructure reflects our propensity to tolerate vulnerabilities and create new risks. Disasters are also responsible for the recent rise in human displacement, which has led to the so-called refugee crisis in Europe and such diverse countries as the United States, Lebanon, and Colombia. In part II, four chapters explore such challenges and the decision-making process required during reconstruction efforts. Chapters 4 and 5 address questions of shelter, debating whether, how long, in what form, and where people should be housed in the immediate aftermath of a crisis. Chapters 6 and 7 take on broader dynamics of participation and disaster aid as part of a disaster risk reduction response. The chapters in part II also address questions of voice (whose voices get heard, how, and with what influence) and trade-offs among possible policy and practice alternatives.

Part III turns to what ought to be. The effects of climate change, increasing inequality, and persistent poverty will exacerbate disasters in years to come. How are we going to respond to these threats and new ones? The debates in part III explore the critical role of politicians, development actors, and disaster managers in making our cities and settlements safer. We also explore the ethical considerations needed to protect ecosystems and ensure the quality of human and nonhuman lives.

Throughout the book, we delve into the controversies that emerge and the difficult decisions that must be made when cities, people, and the environment are at risk. Reducing risks and responding to disasters demand tough decisions and require a comprehensive understanding of socioeconomic, cultural, environmental, and technological conditions. Readers seeking an optimal, one-size-fits-all solution to dealing with risk will be disappointed. As the planet warms and urbanizes at a rapid pace, urban systems are becoming increasingly complex; resources are becoming scarce; inequality, poverty, and other social injustices persist; and complex governance structures have become the norm. Today, policies, projects, and ideas have become difficult to tailor and to implement. Given our evident capacity to destroy nature and seeming incapacity to eliminate social injustices, we are undoubtedly facing some of the toughest choices of human action in recent history. It is time to confront the complex scope and consequences of this challenge to humanity and the world.

NOTES

1. Warren Cornwall, "As Sea Levels Rise, Bangladeshi Islanders Must Decide Between Keeping the Water Out—or Letting It In," *Science*, March 1, 2018.
2. Mohammad Shahidul Hasan Swapan, Md Ashikuzzaman, and Md Sayed Iftekhar, "Equitable Access to Formal Disaster Management Programmes: Experience of Residents of Urban Informal Settlements in Bangladesh," in *Natural Hazards and Disaster Justice*, ed. Anna Lukasiewicz and Claudia Baldwin (Singapore: Springer, 2020): 169–183.
3. Mohammed Sarfaraz Gani Adnan, Anisul Haque, and Jim W. Hall, "Have Coastal Embankments Reduced Flooding in Bangladesh?," *Science of the Total Environment* 682 (2019): 405–416.
4. Jason Cons, "Staging Climate Security: Resilience and Heterodystopia in the Bangladesh Borderlands," *Cultural Anthropology* 33, no. 2 (2018): 266–294.
5. Reiner Grundmann, "Climate Change and Knowledge Politics," *Environmental Politics* 16, no. 3 (2007): 414–432.
6. Gonzalo Lizarralde, *Unnatural Disasters: Why Most Responses to Risk and Climate Change Fail but Some Succeed* (New York: Columbia University Press, 2021).
7. Tapan Dhar, Lisa Bornstein, Gonzalo Lizarralde, and S. M. Nazimuddin, "Risk Perception—A Lens for Understanding Adaptive Behaviour in the Age of Climate Change? Narratives from the Global South," *International Journal of Disaster Risk Reduction* 95 (2023): 103886; Lee Bosher, Ksenia Chmutina, and Dewald van Niekerk, "Stop Going Around in Circles: Towards a Reconceptualisation of Disaster Risk Management Phases," *Disaster Prevention and Management: An International Journal* 30, no. 4/5 (2021): 525–537.
8. Lizarralde, *Unnatural Disasters*; Wesley Cheek and Ksenia Chmutina, "'Building Back Better' Is Neoliberal Post-Disaster Reconstruction," *Disasters* 46, no. 3 (2022): 589–609.

9. United Nations Office for Disaster Risk Reduction (UNDRR), *Sendai Framework for Disaster Risk Reduction 2015–2030*, 2015, https://www.preventionweb.net/files/43291_sendaiframeworkfordrren.pdf; Nibedita S. Ray-Bennett, and Hideyuki Shiroshita, “Disasters, Deaths and the Sendai Framework's Target One: A Case of Systems Failure in Hiroshima Landslide 2014, Japan.” *Disaster Prevention and Management* 28, no. 6 (2019): 764–785.
10. J. C. Gaillard, “Disaster Studies Inside Out,” *Disasters* 43 (2019): S7–S17.
11. World Meteorological Organization (WMO), “Weather-Related Disasters Increase Over Past 50 Years, Causing More Damage but Fewer Deaths,” August 31, 2021, https://wmo.int/media/news/weather-related-disasters-increase-over-past-50-years-causing-more-damage-fewer-deaths.
12. UNDRR and Centre for Research on the Epidemiology of Disasters (CRED), *The Human Cost of Disasters 2000–2019: An Overview of the Last 20 Years 2000–2019*, 2020; Morgan Scoville-Simonds, Hameed Jamali, and Marc Hufty, “The Hazards of Mainstreaming: Climate Change Adaptation Politics in Three Dimensions.” *World Development* 125 (2020): 104683.
13. WMO, “Weather-Related Disasters Increase.”
14. Keith Smith, *Environmental Hazards: Assessing Risk and Reducing Disaster* (London: Routledge, 2013).
15. Gonzalo Lizarralde, Cassidy Johnson, and Colin Davidson, eds. *Rebuilding After Disasters: From Emergency to Sustainability* (London: Taylor & Francis, 2009).
16. Nibedita S. Ray-Bennett and Hideyuki Shiroshita, “Disasters, Deaths and the Sendai Framework's Target One: A Case of Systems Failure in Hiroshima Landslide 2014, Japan.” *Disaster Prevention and Management* 28, no. 6 (2019): 764–785.
17. Scoville-Simonds, Jamali, and Hufty, “The Hazards of Mainstreaming.”
18. Gonzalo Lizarralde et al., “Does Climate Change Cause Disasters? How Citizens, Academics, and Leaders Explain Climate-Related Risk and Disasters in Latin America and the Caribbean,” *International Journal of Disaster Risk Reduction* 58 (2021): 102173.
19. Gaillard, “Disaster Studies Inside Out.”

20. Melissa L. Finucane et al., "Short-Term Solutions to a Long-Term Challenge: Rethinking Disaster Recovery Planning to Reduce Vulnerabilities and Inequities," *International Journal of Environmental Research and Public Health* 17, no. 2 (2020): 482.
21. Intergovernmental Panel on Climate Change (IPCC), "AR6 Synthesis Report: Climate Change 2023," 2023, https://www.ipcc.ch/report/ar6/syr/.
22. David Wachsmuth, Daniel Aldana Cohen, and Hillary Angelo, "Expand the Frontiers of Urban Sustainability," *Nature* 536 (2016): 391–393.
23. Lizarralde, Johnson, and Davidson, *Rebuilding After Disasters*.
24. Arry Retnowati et al., "Environmental Ethics in Local Knowledge Responding to Climate Change: An Understanding of Seasonal Traditional Calendar Pranotomongso and Its Phenology in Karst Area of Gunungkidul, Yogyakarta, Indonesia," *Procedia Environmental Sciences* 20 (2014): 785–794.
25. Vandana Desai, Rob Kitchin, and Nigel Thrift, "Aid," in *International Encyclopedia of Human Geography*, ed. Rob Kitchin and Nigel Thrift, 84–90 (Oxford: Elsevier, 2009).
26. Alexandra Titz, Terry Cannon, and Fred Krüger, "Uncovering 'Community': Challenging an Elusive Concept in Development and Disaster Related Work," *Societies* 8, no. 3 (2018): 71.
27. Alexander Koensler and Cristina Papa, "Introduction: Beyond Anthropocentrism, Changing Practices and the Politics of 'Nature,' " *Journal of Political Ecology* 20, no. 1 (2013): 286–294.
28. Gonzalo Lizarralde, "Investing in Disaster-Risk Consultants and Visibility," in *Investing in Disaster Risk Reduction for Resilience: Design, Methods and Knowledge in the Face of Climate Change*, ed. A. Nuno Martins et al. (New York: Elsevier, 2022).
29. Ben Wisner, "Five Years Beyond Sendai—Can We Get Beyond Frameworks?," *International Journal of Disaster Risk Science* 11, no. 2 (2020): 239–249, 240.
30. Ben Wisner et al., *At Risk: Natural Hazards, People's Vulnerability and Disasters*, 2nd ed. (London: Routledge, 2004).
31. Gaillard, "Disaster Studies Inside Out."
32. The Organisation for Economic Co-operation and Development (OECD) is an international organization working to build better

policies to foster prosperity, equality, opportunity, and well-being for all.

33. J. C. Gaillard and Lori Peek, "Disaster-Zone Research Needs a Code of Conduct," *Nature* 575 (2019): 440–442.
34. Gaillard, "Disaster Studies Inside Out."
35. Gonzalo Lizarralde et al., "We Said, They Said: The Politics of Conceptual Frameworks in Disasters and Climate Change in Colombia and Latin America," *Disaster Prevention and Management* 29, no. 6 (2020).

PART I

LANGUAGE AND MEANINGS

Language and Meanings of Risk, Adaptation, and Resilience in Negril, Jamaica

TAPAN DHAR, GONZALO LIZARRALDE, AND LISA BORNSTEIN

WHO IS KILLING THE CORALS, AND WHO WILL BUY THE BOB MARLEY PORTRAIT?

"Discover coral reefs bursting with color, schools of fish, and vibrant clusters of sponges," states the advertisement for a Long Bay resort.[1] Long Bay is a seven-mile beach of white sand in Negril, a small town of eight thousand residents on Jamaica's west coast. Hundreds of resorts and snorkeling clubs face the sea, with buildings located as close as ten meters from the water (see figure I.1). When we visited Negril in 2015, we saw few snorkelers but substantial construction work—several hotels were in the process of being renovated or extended during our stay.

FIGURE I.1 The seven-mile-long, but increasingly narrow, Long Bay Beach in Negril.

Craft and souvenir shop owners were working on products to sell over the following tourist season.

Antony, a tall and generous resident, has operated a small souvenir shop in Long Bay for more than three decades. When we met him, he voiced a serious concern: "We are losing our beach every year. Ten years ago, the beach was wider, nearly double what you see right now." Hurricanes such as Michelle in 2001 and Wilma in 2005 have caused part of the damage, though Negril has since been spared other major storms. Some government representatives and academics are now blaming global warming as well for transformations to the landscape.

Edward Robinson, an emeritus professor of geology at the University of the West Indies, and his colleagues have studied the combined effects of erosion, global warming, and sea level rise in the region.[2] They anticipate that beach erosion will be up to ten meters by 2030 and may exceed twenty-one meters by 2050. A 2014 report mandated by Jamaica's National Environment and Planning Agency (NEPA) estimates that as many as six kilometers of the beach will erode over the next thirty-seven years. According to the organization, "estimations based on global sea level rise projections and local storm wave predictions suggest that the impact on Negril will be devastation." Conservative projections of sea level rise for 2060, combined with an extreme fifty-year storm return period in Negril, anticipate that "approximately 50 percent of the beach will lose more than half of its present width."[3] Similar problems exist in other islands in the region. A team from Waterloo University estimates that sea level rise might partially or fully damage 29 percent of coastal resorts in the Caribbean basin.[4]

According to the NEPA report, indirect negative impacts of sea level rise in Negril include "the potential decrease in tourists . . . , potential loss in infrastructure and investments, and increased unemployment."[5] Many locals agree. Antony is worried that fewer tourists will visit his shop; as he put it, "who will purchase this bag or this Bob Marley portrait; and what will we do then?" Nearly 82 percent of Jamaican employment and 70 percent of Jamaica's production activities are concentrated in coastal areas.[6] Negril's economy depends on tourism, which constitutes more than 5 percent of the national economy. It is the third most visited resort area in Jamaica and provides the second highest number of direct jobs in the country.[7]

But Antony is not only concerned about a narrow beach. "Our reefs are dying," he adds. "More than half of them are now pale or colorless." Over the past three decades, studies have confirmed that Antony is correct. Even as early as 1991, scientific reports warned that "many corals [in Negril] may die within a few years unless current threats to the reef are abated."[8] The 2023 IPCC also predicted with "very high confidence" that reefs in the region could disappear in the near future.[9]

Coral bleaching in the Caribbean has often been associated with global warming, but climate change is not solely to blame for Negril's problems.[10] Human actions are also responsible. Water pollution is linked to small industries and manufacturing facilities spilling chemicals and oil into the sea.[11] The development of the town is marred by a chronic lack of planning. Regulations are rarely enforced. Antony explained to us: "We have polluted seawater the most. Many boat-repairing workshops are located along the south Negril River. They emit toxic chemicals in the water that flows into the sea and impact seagrass and reefs."

A risk and vulnerability study conducted in 2014 found connections between beach erosion and the destruction of coral reefs and marine ecosystems. The study discovered that seagrass beds and nearshore coral reefs "play a critical role in stabilization of the shoreline."[12] According to these researchers, the absence of reefs in some areas causes about 80 percent of shoreline erosion, and the absence of seagrass accounts for 41 percent.

These ecological relationships raise a difficult question: What is killing the marine ecosystem and putting the livelihoods of Antony and thousands of Jamaicans at risk? Whereas some experts point to long-term global effects, others find more immediate human causes at play. This controversy was already heightened when it was superseded by another conflict in 2014.

BREAKING THE BREAKWATERS

In 2014, the Jamaican government launched an ambitious project to fight beach erosion. The ten-million-dollar project consisted of building two breakwater systems. Breakwaters are often concrete structures designed to reduce the impact of waves, thus preventing sand from being washed away and protecting beaches from erosion. The project was part of the Government of Jamaica's Adaptation Fund Program aimed at enhancing the resilience of coastal areas and protecting local livelihoods.[13] NEPA, the organization in charge of the project, designed the breakwater solution based on climate change projections and simulations of different scenarios.

When we visited Negril in 2015, we witnessed a conflict among citizen groups, hotel owners, and NEPA concerning the secondary effects of the breakwater project. Opponents argued that the government plan was not the right solution to the real problem. The tension received national attention. A letter to the *Jamaica Observer* asserted "Breakwater Project Will Break Negril."[14] Headlines in the *Gleaner* stated "Ignoring the Critics! NEPA Pushing on with Negril Breakwater Despite Criticisms" and "Government, Hoteliers Still Divided Over Negril Breakwaters."[15]

Antony thought the government project was absurd, noting that "we are killing our reefs; we are polluting our beaches. The breakwater project would ruin everything." Hoteliers also wanted to halt the project, fearing its implementation would damage Negril's tourism industry.[16] Joseph Brown, a hotel manager and member of the Jamaica Hotel and Tourist Association, was concerned about the project's negative impacts. He agreed that human activities and lack of law enforcement damaged ecosystems and tourism, arguing that the project would cause more erosion if not properly designed and implemented. "It would be a big disaster," he concluded.

NEPA's studies anticipated several negative impacts of the project, including the "destruction of natural marine habitats from the laying of boulders on the seafloor" and risks of "habitat fragmentation between the seagrass beds in the lagoon and surrounding reefs."[17] More surprisingly, NEPA seemed to recognize that the structures could be poorly designed and potentially dangerous, referring to the "visual impact of above-water portions of breakwater structures" and an "increased accident potential for marine vessels running aground the breakwaters."[18] But NEPA's 2014 report provided answers to almost all the problems it uncovered. The report concluded that "vegetating the sections of the breakwaters that are above water could make the breakwaters more aesthetically pleasing, giving them the appearance of offshore island cays" before acknowledging that this vegetation "could be easily washed away during severe wave climate conditions."[19]

At one point, NEPA's chief executive officer became particularly defensive, complaining in a press release of a "campaign by selected stakeholders from Negril designed to mislead and confuse the public and, at times, denigrate the evidence-based and science-based decision arrived at by the Natural Resources Conservation Authority and the National Environment and Planning Agency."[20]

Were locals ignoring science, as NEPA suggested, or were scientists and technocrats ignoring locals, as villagers argued? How did NEPA navigate this debate and simultaneously acknowledge its internal contradictions and project risks?

Wider narratives of resilience and adaptation were useful to NEPA.[21] NEPA promoted the breakwater project as a strategy to "adapt Negril to climate change effects" and "achieve resilience in Long-Bay."[22] It framed the problem as a global one, linking it to changes in the atmosphere and water temperature and advancing new infrastructure as a "fix."[23] Breakwater systems could mitigate risk, allowing humans and ecosystems to coexist with natural hazards. These narratives masked the role of pollution, lack of regulation, and poor law enforcement. By focusing on the role of global warming, authorities shifted attention from solving local problems to justifying new investments.

In July 2016, under severe pressure from citizens, businesspeople, and social groups, NEPA announced that it was abandoning the breakwater project.[24] Funding and aid agencies moved to promote climate adaptation in other places in Jamaica. A few years later, the country approved the Comprehensive Disaster Risk Management Policy and Strategy (2020–2040), a policy that resulted from a long negotiation between Jamaica and the Inter-American Development Bank and that allows Jamaica to access $285 million to alleviate the financial impact of climate change effects.[25]

The breakwater project did not survive, but the dialogue it raised on adaptation, resilience, and sustainable infrastructure persisted. These concepts entered local conversations and are now familiar to Antony and other residents. These terms still mean different things to locals, politicians, scholars, and NGOs, but they are there to stay.

WHY TERMS AND MEANINGS MATTER

Scholars have found that, until about a decade ago, most disaster risk reduction policies in Latin America and the Caribbean did not include concepts such as resilience, adaptation, and adaptive capacities.[26] In Cuba, for example, the United Nations Development Program (UNDP) started collaborating with local institutions after 2012's Hurricane Sandy. At that time, the government's stated intent was to protect citizens. By 2015, UNDP, UN-Habitat, the Cuban National Planning Institute, and the

Cuban Civil Defense were collaborating on a project aimed at "enhancing urban resilience in Cuba," an initiative supported by the United Nations that produced a series of guidelines and publications on how to adopt a resilience framework at a municipal level.[27] The language of resilience, adaptation, and sustainability is now common in Colombia, Ecuador, Chile, and other countries in the region.[28]

As suggested in the Jamaican cases above, the words employed, and the way issues are framed, matter. The chapters in this section explore why and how. In chapter 1, Thomas Fisher and Michael Mehaffy debate the scope of urban vulnerability, challenge notions of fragility, and analyze the way we describe cities in an age of global warming. Today, our screens are full of news of urban disasters: floods, tsunamis, earthquakes, tornadoes, pandemics, and wildfires. But are cities and human systems really more fragile today than a few decades ago? The authors observe that cities have helped reduce some forms of vulnerability and exacerbated others. For example, our urban areas concentrate economic opportunities and reduce transportation needs, but they also bring together large numbers of people who can be affected by a dangerous event. The debate addresses long-term trajectories of progress, decline, and crisis, as well as differing conceptions of resilience. As such, the debate shows how the framing and definition of urban problems shapes subsequent actions. The chapter concludes with observations that as the world warms and urbanizes, and as people become increasingly connected, environmentally conscious, and dependent on technology, perceptions of imminent disaster are more common. We are more aware that disasters will occur but still underestimate the chances that they might happen in our cities.

In chapter 2, J. C. Gaillard and Carmen Mendoza-Arroyo discuss neocolonialism in disaster studies and risk reduction work. They ask, to what extent do foreign narratives, concepts, and ideas derived from Northern contexts influence action in the Global South? People in the Global South suffer most from disasters and the effects of climate change, but their tragedies and struggles are often explained through concepts developed in wealthy nations. This debate unpacks academic colonialism in disaster studies and explores whether concepts and ideas derived from the Global North dangerously dominate disaster scholarship and practices in the Global South. They further question whether this level of influence generates problematic dependencies, distorted understandings of conditions, or other impediments to responses to the many risks facing us. Issues of epistemology, North-South power relations, and the positionality of researchers and disaster specialists are raised. The chapter's authors, online commentators, and moderator agree that various foreign influences inform interventions that are economically and culturally unfit to local dynamics and thereby fail to address root causes of vulnerability. However, classifying researchers and studies according to geographic locations and origins is often challenging. Offered instead is an ethical perspective on disaster studies, with practices articulated around notions of social justice and based on a thorough understanding of, and respect for, local knowledge and values.

Resilience is perhaps the most popular and ambiguous concept used in contemporary disaster research, communication, and policy. This concept has been adopted in different areas of expertise, from geography to political science to architecture. But is resilience useful in the field of disaster risk reduction, or is it simply a new buzzword? In chapter 3, Daniel Aldrich and Jonathan Joseph explore the advantages and disadvantages of the concept of resilience, particularly as applied to the design of the built environment for disaster risk reduction. How useful is the concept? To what extent does it add value to disaster research and work, whether in Negril or in other places? Which ideas and attitudes are masked, and which are revealed, when we argue that a place such as Long Bay is or should be resilient? How does resilience help us understand the fragile relationships among built, natural, and social environments? The debaters in this chapter explore how resilience is understood, promulgated, and employed: is it a buzzword? a helpful term? or a theory that can move us dramatically forward in risk reduction, response, and recovery? Participants in this debate agree that resilience is a problematic concept but, perhaps surprisingly, refuse to abandon it as a way to understand disaster risk and responses. The debate leads to a summary observation that practical and conceptual frameworks require thorough assessment of power relationships between stakeholders, forms of domination, and local politics of disaster risk creation and reduction.

Throughout the chapters in part I, we show that common notions in climate change are insufficiently challenged in disaster studies and practices, which leads scholars and practitioners to perceive a consensus in the field that does not exist. Residents and authorities in Negril share a desire to reduce the impacts of hazards, but they agree neither on the status and definition of these hazards nor on how they should be addressed. Local narratives and explanations of risk, such as those proffered by Antony, remain overlooked by decision-makers and scholars. The debates here suggest that more attention be placed on the meaning of terms deployed to explain risk, vulnerability, and destruction. The language of risk brings its own risks.

NOTES

1. See Skylark Negril Beach Resort, "You Need to See These 5 Spots for Snorkeling in Negril, Jamaica," accessed March 12, 2023, https://skylarknegril.com/blog/snorkeling-in-negril-jamaica/.

2. Edward Robinson and Malcolm Hendry, "Coastal Change and Evolution at Negril, Jamaica: A Geological Perspective," *Caribbean Journal of Earth Science* 43 (2012): 3–9.
3. National Environment and Planning Agency (NEPA), *Environmental Impact Assessment: Construction of Two Breakwaters at Long Bay, Negril, Westmoreland*, 2014, https://www.nwa.gov.jm/sites/default/files/publications/Negril%20Breakwaters%20EIA%20Report_0.pdf, 8.
4. Daniel Scott, Murray Charles Simpson, and Ryan Sim, "The Vulnerability of Caribbean Coastal Tourism to Scenarios of Climate Change Related Sea Level Rise," *Journal of Sustainable Tourism* 20, no. 6 (2012): 883–898.
5. NEPA, *Environmental Impact Assessment*, 405.
6. Amani Ishemo, "Vulnerability of Coastal Urban Settlements in Jamaica," *Management of Environmental Quality: An International Journal* 20, no. 4 (2009): 451–459.
7. Albert Ferguson, "Negril Wants Government Help to Protect Tourism Product," *Gleaner* (Jamaica, WI), February 24, 2023, https://jamaica-gleaner.com/article/lead-stories/20230224/negril-wants-government-help-protect-tourism-product.
8. Thomas J. Goreau, *Coral Reef Health in the Negril Area: Survey and Recommendation*, 1991, https://www.globalcoral.org/_oldgcra/coral_reef_health_in_the_negril.htm.
9. Intergovernmental Panel on Climate Change (IPCC), "AR6 Synthesis Report: Climate Change 2023," 2023, https://www.ipcc.ch/report/ar6/syr/.
10. Jamie K Reaser, Rafe Pomerance, and Peter O Thomas, "Coral Bleaching and Global Climate Change: Scientific Findings and Policy Recommendations," *Conservation Biology* 14, no. 5 (2000): 1500–1511.
11. Tapan Dhar, "Identifying Climate Change Vulnerability and Adaptation Challenges in the Caribbean SIDS: An Urban Morphological Approach," in *Small Island Developing States: Vulnerability and Resilience Under Climate Change*, ed. S. Moncada et al. (Cham, Switzerland: Springer, 2021), 329–350.
12. NEPA, *Environmental Impact Assessment*, 240.
13. "Negril Breakwater Project Approved," *Gleaner* (Jamaica, WI), December 16, 2014, https://jamaica-gleaner.com/article/news/20141216/negril-breakwater-project-approved.

14. Jane Issa, "Breakwater Project Will Break Negril," *Jamaica Observer*, January 6, 2015, https://www.jamaicaobserver.com/letters/breakwater-project-will-break-negril/.
15. Paul Williams, "Ignoring the Critics! NEPA Pushing on with Negril Breakwater Despite Criticisms," *Gleaner* (Jamaica, WI), May 16, 2015, https://jamaica-gleaner.com/article/news/20150517/ignoring-critics-nepa-pushing-negril-breakwater-despite-criticisms; Petre Williams-Raynor, "Government, Hoteliers Still Divided Over Negril Breakwaters," *Gleaner* (Jamaica, WI), June 15, 2015, https://jamaica-gleaner.com/article/lead-stories/20150616/government-hoteliers-still-divided-over-negril-breakwaters.
16. Zadie Neufville, "Row Over Jamaica's Bid to Slow Beach Erosion," *Caribbean Life*, February 6, 2015, https://www.caribbeanlife.com/row-over-jamaicas-bid-to-slow-beach-erosion/.
17. NEPA, *Environmental Impact Assessment*, 406.
18. NEPA, *Environmental Impact Assessment*, 406.
19. NEPA, *Environmental Impact Assessment*, 410.
20. Peter Knight, "Breakwater Will Build Resilience in Negril," January 9, 2015, https://jamaica-gleaner.com/article/letters/20150110/breakwater-will-build-resilience-negril.
21. Tapan Dhar and Luna Khirfan, "Community-Based Adaptation Through Ecological Design: Lessons from Negril, Jamaica," *Journal of Urban Design* 21, no. 2 (2016): 234–255.
22. Knight, "Breakwater Will Build Resilience in Negril."
23. Christopher Serju, "Negril Beach Erosion Grabs Vaz's Attention." *Gleaner* (Jamaica, WI), November 25, 2019, https://jamaica-gleaner.com/article/news/20191125/negril-beach-erosion-grabs-vazs-attention.
24. Livern Barrett, "Scrapped—Gov't Pulls Plug on Controversial $1-Billion Negril Breakwater Project," *Gleaner* (Jamaica, WI), July 12, 2016, http://jamaica-gleaner.com/article/lead-stories/20160713/scrapped-govt-pulls-plug-controversial-1-billion-negril-breakwater.
25. "Jamaica's Disaster Risk Management Capacity Being Strengthened," *Gleaner* (Jamaica, WI), July 20, 2022, https://jamaica-gleaner.com/article/news/20220720/jamaicas-disaster-risk-management-capacity-being-strengthened.

26. Ernesto Aragon-Duran et al., "The Language of Risk and the Risk of Language: Mismatches in Risk Response in Cuban Coastal Villages," *International Journal of Disaster Risk Reduction* 50 (2020): 101712.
27. A. Muniz Gonzalez, "Plan espacial para la reducción de riesgos y vulnerabilidades ante desastres," PNUD (UNDP), Cuba, 2015; Aragon-Duran et al., "The Language of Risk and the Risk of Language."
28. ADAPTO, *Artefacts of Disaster Risk Reduction: Community-Based Initiatives to Face Climate Change in Latin America and the Caribbean* (Montreal: Oeuvre Durable, 2022).

1

ON FRAGILITY

Are Cities and Human Systems Increasingly Fragile in the Face of Disasters?

WITH THOMAS FISHER AND MICHAEL MEHAFFY

Scholars and practitioners interested in disaster risk reduction often claim that in an urbanized and warming world, human systems—such as those used for transportation, communication, and delivering public services—are increasingly at risk.[1] For them, human progress in its present form endangers ecosystems, wildlife, the atmosphere, and ultimately human life. They often note that technology makes humans dependent on energy, especially carbon fuels. Communication technologies and artificial intelligence pose new risks, as people increasingly depend on computerized systems prone to failure and disruptions. Nuclear and biological war continue to be threats, and capitalist economic systems are unsustainable, especially for the most vulnerable. Overconsumption, fueled by unbridled capitalism, is on the rise.

Cities are often blamed for exacerbating these risks.[2] Cities occupy 3 percent of the earth's surface but are responsible for producing 75 percent of CO_2 emissions.[3] Cities also exacerbate exclusion and deepen the divide between rich and poor, as well as between those with access to technology and those without.

Urban sprawl in metropolitan areas increases commuting times, reducing the quality of life for millions of urbanites. Freedom of movement, even in a so-called globalized world, is a privilege for a small minority. As the world urbanizes, glaciers melt, and oceans warm, more disasters occur. According to this perspective, human systems are increasingly fragile—and perhaps on the brink of collapse.

Other scholars claim that the world has never been more resilient or sustainable than it is now.[4] They note that human progress is real and measurable: in most countries, life expectancy has significantly increased, illiteracy and crime rates have dropped, and there are fewer mortal diseases, wars, armed conflicts, and human rights violations than ever before.[5] For these scholars, the decline of totalitarian regimes and the proliferation of capitalist economies are unquestionable generators of wealth, leading to ongoing decreases in poverty, undernourishment, and famines. Technology has made work, construction, travel, and communication easier and safer.[6]

From this perspective, cities signify one of the greatest steps on the path toward progress. Cities are inclusive, dynamic, and complex structures that connect people; enhance entrepreneurship, culture, and creativity; and create opportunities for prosperity, learning, and entertainment. More importantly, cities bring people together in concentrated areas, reducing the sprawl of human settlements on the planet. Even though cities are increasingly affected by natural hazards, the impact of these events on deaths and injuries per capita is decreasing. Finally, with more advanced technology, disasters can be avoided or mitigated. In sum, this scholarly perspective holds that human systems are far from collapse in the contemporary world and in fact are becoming increasingly resilient.

THOMAS FISHER: CITIES AND HUMAN SYSTEMS ARE INCREASINGLY FRAGILE

Historically, cities have been resilient in the face of disasters, both natural and human-caused, as evidenced by the number of cities worldwide that have been destroyed and eventually rebuilt, often because they occupy strategic locations. The same might be said of social systems. While human populations have faced catastrophes of various kinds, societies have often managed to survive such events, despite loss of life and social disruption.

A historical record of social resilience in the face of disasters, however, should not make us complacent. Two discernible trends may make the human ability to recover from disasters more difficult in the future. People have some control over both trends, and thus some ability to change trajectories. First, while the number of wars and casualties has decreased globally—with obvious exceptions in some places around the world—the frequency and severity of weather-related disasters have increased dramatically in recent decades.[7] Although some still deny this reality, human activity has substantially driven climate change.[8]

The second trend, related to the first, will make it much harder for us to adapt to and recover from increasingly extreme weather events. The burning of fossil fuels has altered climate patterns and also made us increasingly dependent on the global trade of goods to meet our basic needs, increasingly isolated because of our ability to live apart from each other, and increasingly helpless in our ability to live without access to the infrastructural, governmental, and institutional systems that support us.

This helplessness to confront climate events is ironic. Humans have never had more power to bend nature to their will. They have never had more of an impact on the planet and its ecosystems. But as is the case with any dominant species in an ecosystem, the moment it reigns supreme also marks its most likely moment of collapse, as it comes to depend on an ecosystem whose health it has undermined through its own dominance. Humanity now occupies that position in the global ecosystem: we have never been more dominant, powerful, and vulnerable than we are now.

What will bring us down is not what most people think. We can recover from coastal flooding or inland drought, yet we remain vulnerable to threats beyond our perception, such as zoonotic diseases, spread through transcontinental travel, for which we have no immunity or vaccine. Some of our greatest vulnerabilities are enabled by one of our greatest technological achievements: jet aircraft, facilitated by the burning of fossil fuels. How might we avoid such a fate? We should begin by living more sustainably: living frugally, working locally, and traveling as little as possible.

MICHAEL MEHAFFY: CITIES AND HUMAN SYSTEMS ARE NOT INCREASINGLY FRAGILE

I see an instructive paradox at the heart of this question. Many technological systems are becoming more fragile through an increasing reliance on long supply chains, interdependent components, and unstable, unsustainable resource inputs. Yet other human systems—especially urban systems—are becoming more resilient, in large part because societies are gradually learning how to tap the inherent resilience of complex adaptive systems, including cities and related human systems.

To be clear, I am not a techno-optimist. I do not believe that technological innovations can resolve our most intractable problems: resource depletion, ecological destruction, contamination, climate change, vulnerability to catastrophes, and the more subtle but no less worrisome declines in cultural systems. But I do believe that we have the means to address these and other challenges and that we are beginning to do so. One of the most powerful resources available is the inherent capacity of human systems, including cities, to form self-organizing, problem-solving networks that can effectively respond to shocks. This capability is the essence of resilience.

The resilience theorist C. S. Holling famously distinguished between "engineered resilience" and "ecological resilience."[9] He claimed that the former works well to cope with events that remain within the limited specifications for which they are engineered. But many events fall outside this range, especially those that occur as the result of unintended consequences. For example, managed retreat, a popular disaster mitigation approach that relocates people from a disaster-prone area to a relatively safer place, may have the unintended consequence of disrupting incomes and sociocultural practices. Unintended consequences fall beyond expectations; they can even be "far from equilibrium," meaning that it may not be possible to predict, let alone engineer, them. In such a case, the "ecological resilience" conferred by evolutionary and self-organizing processes is useful.

These processes rely on several crucial characteristics, including web-network patterns with redundant connections, fine-grained adaptivity, and the ability to learn and build on previous solutions. Their structure gives them a greater ability to cope with far-from-equilibrium phenomena. For a human system like a city, ecological resilience demands more spatial connectivity of people and resources (e.g., within streets and public spaces), a fine grain of buildings and other adaptable structures, and the capacity to share and build on problem-solving knowledge. For example, the sociologist Eric Klinenberg found that during the Chicago heat wave of 1995, neighborhoods with high rates of social connectivity and "social capital," loosely defined as the benefits conferred from well-connected social networks, experienced much higher survival rates than those in which residents were more isolated.[10] In such cases, urban network connectivity can be a matter of life and death.

Evidence suggests that self-organizing, problem-solving dynamics against disasters work strongly in humans' favor. Disasters like 9/11, the Fukushima tsunami and nuclear accident, and recent earthquakes and storm events show this remarkable dynamic of resilience in action, as people recover and rebuild. Partly because of such resilience, we are on average safer, healthier, more peaceful, and certainly more prosperous today than at any other time in the history of our species.

To be sure, we have much work to do. Too many of our systems are still fragmented and poorly networked, and they lack the kind of coherent feedback required for ecological resilience and sustainability—in particular, feedback for externality impacts. Urban sprawl, for example, is the physical manifestation of this dangerous form of disorder. But there are other forms too, including regulatory, technological, and economic systems. We remain too dependent on an economy of depletion, and we have not yet come to terms with a pressing transition to an economy of repletion. Yet the means to make this transition are available in both natural resilience and "natural urbanism," which constitute powerful forces for a better future.

THOMAS FISHER'S REBUTTAL REMARKS

Human societies, and especially cities, have a long history of resilience, as argued by my colleague Michael Mehaffy. Indeed, cities have rebounded after experiencing almost complete desertion, whether because of warfare, in the case of the Roman destruction of Carthage, now part of the city of Tunis, or due to environmental collapse, in the case of the Native American abandonment of Cahokia, now part of the St. Louis metro area. The resettlement of such places happens for a couple of reasons. Cities usually occupy strategic locations that prompt their rebuilding even after a catastrophic event. And, as Mehaffy argues, cities also constitute complex adaptive systems in which networks of people respond to and bounce back from shocks.

The time scale in which such a rebound happens matters significantly. To bounce back quickly, as Berlin did after World War II, is one thing; taking hundreds or even thousands of years, in the case of Carthage or Cahokia, is quite another. While human communities may return to key locations after the societies that once occupied destroyed sites have disappeared, we cannot call this phenomenon resilience. According to the ecologist C. S. Holling, it represents a collapse, as constantly happens in an ecosystem. His theory suggests that humanity is subject to the same panarchic cycle as every other species that dominates an environment, to the extent that it destroys the ecosystem it depends on and faces extinction—or at least major disruption—in the process.[11]

Mehaffy rightly points out that cities and their inhabitants need spatial connectivity, and he is correct in arguing that knowing our neighbors can help us survive short-term threats such as heat waves. But one of the greatest threats we face is our own complacency about larger dangers. Human societies have survived collapses in the past, typically in remote places like Easter Island or Greenland, as shown by the environmental historian Jared Diamond.[12] Never in human history, though, have we faced the possible collapse of the global human ecosystem, in part because of the spatial connectivity lauded by Mehaffy. Every other species on the planet occupies relatively small ecosystem patches that collapse and reorganize without disrupting the rest, but our species has engaged in an experiment over the past few centuries of creating a single global economy so interconnected and codependent that a failure of any part can bring down the whole.

PARTICIPANT'S COMMENTS

Miguel Escobar, Principal Architect and Planner, Future Cities Group Montreal, Canada

Cities are becoming increasingly fragile. As systems become more sophisticated, additional resources, and thus revenues, will be required. Cities will in turn become exponentially more expensive to live in. Eventually, there will be a breaking point where the haves and the have-nots will come to loggerheads. Populist ideals will confront capitalist norms. And no matter how high-tech and futuristic a city becomes, all democratic societies are subject to rapid and catastrophic policy shifts—whether by internal forces such as the disastrous downturn of a modern nation like Venezuela or by external forces such as the destruction of modern cities in Syria. Even policy shifts in bastions of democracy such as the United States can be catastrophic. Take, for instance, the Trump administration's

> response to Hurricane Maria in Puerto Rico. Under any other administration, whether Democrat or Republican, the response would have been much more sympathetic. If the Trump administration continues with these policies, American cities will not only become increasingly fragile but will approach a tipping point. The constitution and government of the United States are based on many layers of checks and balances that make it resilient. Cities too must bolster their constitutions and regulations so as to ensure that a one-term administration does not undo the work of an entire society.

This may sound pessimistic and hopeless, but it is not. Humanity thrived for most of our history as a species in ecosystem patches like every other animal, and we need to do so again if we want to survive in the future. The size of a patch does not matter; what matters is the self-sufficiency and self-reliance of the people living there. Every place needs to begin imagining how it would survive a global collapse, how it would thrive off the grid: feeding, housing, and employing its people with the renewable resources it has at hand—something that humans once did very well.

MICHAEL MEHAFFY'S REBUTTAL REMARKS

I agree with Thomas Fisher on a number of points: first, that cities have historically been resilient, as have social systems; second, that we have more recently set in motion a number of destructive trends, including climate change; and third, that we have increased our vulnerability as the result of progressive reliance on long supply chains, interdependent components, and unstable, unsustainable resource inputs, as stated in my opening remarks. I also agree with his more optimistic assertion that we can recover reasonably well from disasters like coastal flooding and inland drought.

So where do we disagree? I would say that we differ in the conclusions we draw from the same evidence. Fisher sees an inevitable vulnerability originating in the fact of human dominance. He considers, especially, that zoonotic diseases transmitted from animals, like HIV and Ebola, and the threat that they will be spread by air travel "will bring us down." This claim strikes me as overly speculative. I do not doubt that such diseases are a major threat, but I question whether this danger differs from that of past plague pandemics, from which societies recovered quickly and thereafter even prospered. In other words, life is a dangerous business, but overall, we have become skilled at managing its downsides.

I confess that I am also skeptical when it comes to specific prognostications of fragility in the face of disaster. History has a way of playing out in surprising ways, in part because we are not victims of fate but can perceive and adapt to real dangers. The so-called Y2K disaster of the late 1990s, when it was widely predicted that we would see massive systems failures in the year 2000 as the result of our data systems being calibrated to two-digit years, is a case in point. The disaster did not happen largely because we recognized the real danger and reacted to it by adapting our systems.

Fisher also makes the case that we are now the dominant species in our ecosystem and that this very dominance makes a species "the most liable to collapse." I do not find this idea persuasive because I can think of many counterexamples. Dinosaurs, for example, were the dominant species for millions of years. In the end, it was an external event (a massive meteor strike), not their dominance, that destroyed them.[13]

PARTICIPANT'S COMMENTS

Mauro Cossu, PhD candidate, Faculty of Environmental Design, Université de Montréal, Canada

I find that the question is not balanced, thus somehow influencing the discussion. While the concept of fragility is fairly intuitive, it is not as easy to understand its opposite. We refer constantly to resilience, which is not actually the opposite of fragility. Moreover, since criteria and indicators to measure, quantify, and visualize fragility have not been established, it seems that the positions taken are based mostly on perceptions. At any scale, fragility is a consequence of the accumulation of risks, or the cumulative effects of multiple risks, that result in a greater likelihood and intensity of urban vulnerability to disasters.

We can consider the city as a complex adaptive system that is composed, in turn, of subsystems and elements that interact with each other. In a more empirical sense, the city is composed of neighborhoods, communities, and individuals that are all part of a larger system, but that do not share the same degree of fragility. In fact, as systems become complex, they become less fragile and therefore less likely to interrupt their operations in the case of a major disturbing event. Fragility—unlike resilience—increases in simpler subsystems such as neighborhoods and local communities, such that an individual's fragility is higher than a social group's, which is more fragile than the city as a whole.

I agree with Fisher that we must not be complacent, and I do not mean to minimize the grave dangers we face. But I believe that we have the inherent and growing capacity to generate resilience, both in our cities and in other systems. They are not inert forces that operate independently of human choice: we created them, and we continue to shape them daily. We can self-organize and generate resilience through wiser actions. We can bestow as much resilience on these systems as on natural systems if we so choose.

THOMAS FISHER'S CLOSING REMARKS

I agree with Michael Mehaffy that we have the capacity to create more resilient systems and cities through wiser actions, should we choose. However, this raises the question as to the wiser actions we should take and the choices we should make. To address these issues, we might first consider the feature that has left our species so vulnerable to catastrophic events, often of our own making: a dominant mindset in which we see ourselves as exceptional in the natural world and smarter than other animals. This perspective, apart from its basic stupidity, has encumbered our understanding of other species, which largely demonstrate levels of resilience far greater than ours in the face of climate and ecosystem disruptions caused by our carelessness and selfishness. So yes, we have the capacity to create more resilient cities and systems, but not until we choose to learn from species much older and wiser than humankind.

What are some of the lessons we might learn? First, that resilience has little to do with technology. Our reliance upon engineered solutions to external threats—from expensive public works projects to exhaustive public health interventions—only shows how little knowledge we have gathered from other animals, whose resilience depends on adaptive behaviors that largely involve retreating from or uniting in the face of a threat or dispersing in order to minimize its impact. In human terms, this approach would mean settling inland away from coasts prone to flooding, building natural barriers that can withstand extreme events, creating shelter that can support an entire community, and distributing into small groups that could be sustained for a lengthy period of time.

This first lesson leads to a second one: resilience primarily involves collective creativity and social innovation. The individualism arising from the myth of genius and independence thus represents one of the greatest threats to our species. Animals that split from their pack and do not connect to a new group rarely survive long; humanity has begun to experience this phenomenon as a species, as we have separated ourselves from the ecosystems on which we depend. Instead of our current obsession with creative genius, we need to instill collaborative creative skills, relearn the experimental abilities that once helped us thrive, and embrace diversity not as a box to check but as a source of new ideas and cultural adaptations to learn from and implement.

Humans have the capacity for more humility in the face of disruptions and for more openness to learning from others, human and nonhuman alike. The design community likewise has a responsibility to help people envision a new way of inhabiting the planet. The irony implied by this design imperative is that humans lived sustainably for most of their existence as a species, but we must relearn what we once knew if we hope to have any future at all.

MICHAEL MEHAFFY'S CLOSING REMARKS

This discussion has revealed for me several interesting points on which Thomas Fisher and I agree and disagree. We agree that this moment is not a time for complacency and that we all face big challenges ahead to manage resource depletion, ecological destruction, contamination, and, as a result of these three issues, the grave threat of climate change.

I have argued that we are not passive observers of systems that are resilient or not, but rather that our own actions or inactions form integral components of the behavior of a system. The questions raised in the context of cities under risk are what kinds of actions we will take, and how urban structures facilitate or obstruct those actions.

I previously noted the importance of the connectivity of urban networks in disasters like the 1995 Chicago heat wave, which was literally a matter of life and death for many. Since urban network connectivity is increasing, I see a hopeful trend for urban resilience and for the choices we can—and I stress the word *can*—take. For me, this trend is positive: we are not powerless in the face of inexorable forces.

This last point may be where Thomas Fisher and I disagree most significantly. He sees large-scale network connectivity, especially the networked global economy, as a threat. He therefore recommends a retreat into what he refers to as "ecosystem patches": smaller and more autonomous regional units.

But I think he misidentifies the threat to cities and human systems. After all, networks of global trade have existed for many centuries. Global networking is not in itself a threat to resilience and can, in fact, provide resources in times of need. For example, countries around the world marshaled their resources to assist with recovery following the 2005 Asian tsunami and the 2010 Haitian earthquake. Greater isolation in such an event would only compound disaster.

Fisher helpfully referenced Jared Diamond's 2005 book *Collapse*, which is a case study of twelve societies, eight of which collapsed and four of which avoided collapse.[14] The key difference between these categories was not the degree to which the societies were inter-networked with others; in fact, Easter Island, which did collapse, was quite isolated. Instead, the difference was how societies chose to create and regulate their networks, especially networks of feedback. Societies that did not have effective network connectivity between actions and their downstream consequences—so-called externalities—tended to experience a "tragedy of the commons," the destruction of long-term common interests by those reaping short-term rewards.[15] To me, this is the most profound threat we face today, and now on a global scale.

PARTICIPANT'S COMMENTS

Manas Murthy, PhD candidate, Department of Architecture, University of Oregon

This debate has referenced facts, indicators, and evidence, and I do believe these are crucial resources for backing up arguments. However, these are extremely fraught sources. I refer to Paul Dourish's work on data-driven urbanism and the social science of informatics to ground my critique. The "data-driven epiphany" discussed by Pinker is infectious. His transition from Blank Slate to Enlightenment invokes the power of optimism, even heroism: we will rally behind

> his call to arms, leaving behind our hopeless, numbing pessimism. The rhetoric of progress as it is imbued with the rhetoric of reason and data objectivity will bring us out of a cynical funk and put us to collective work toward the betterment of humanity. However, Pinker also warns us against the fatalism and radicalism that might be expected outcomes of an alarmist narrative. In this case, panic and fear might not be productive motivators for positive action. I believe that while Fisher and Mehaffy have presented lucid and cogent arguments to support their speculations and projections for the future, it is more crucial to understand that both wield a coercive power to incite action—neither adopts a fatalist view. This is the best outcome of the debate, because it emphasizes a consequentialist belief in the cocreation of our future and the relevance of action in that direction.

I think the real question, then, is not whether we have global networks, but whether global networks—or networks at any scale—are structured to increase feedback and resilience. As I have argued, we now have much better knowledge about how to achieve these outcomes—and that is a promising advancement.

THE MODERATOR'S CLOSING REMARKS: ARE CITIES BECOMING INCREASINGLY FRAGILE? THE ETHICS OF A NARRATIVE OF PROGRESS

According to most data, cities, communication systems, and transportation networks are at lower risk of destruction today than decades ago. But about 75 percent of participants in our latest online debate consistently argued the opposite. As the world warms and people become increasingly connected, environmentally conscious, and technology-dependent, their perception of imminent disaster looms heavy.

Why do we find it easier to believe that cities are becoming increasingly fragile, even when evidence suggests otherwise? Some scholars blame the media. Steven Pinker, a strong believer in human progress, reminds us that media sources often report on sudden and morose events rather than on positive slow changes. Other scholars point to the politics of fear. The

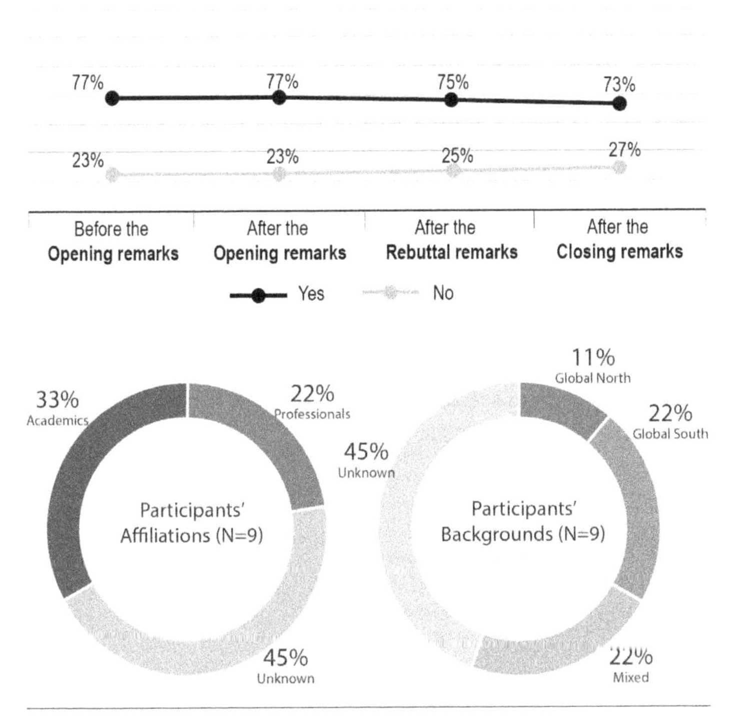

FIGURE 1.1 (*Top*) Results of votes at each stage of the debate. (*Bottom*) Distribution of affiliations and backgrounds among participants in the discussion.

perceived risk of tragedy is a powerful tool for crafting policies guided by ideology. For instance, fears of mass migration and the influx of refugees are now common in Western countries.[16] But such "influxes" represent only a fraction of forcibly displaced populations worldwide. Most migrants go to poor countries in the Middle East, Latin America, and Africa. The widespread belief that industrialized nations are the main recipients of refugees may legitimize "humanitarian" businesses and generate votes in Europe and the United States, but this is a fundamentally inaccurate phenomenon.

In many cases, diverging interpretations of data shape what constitutes a "fact" or "concept." Consider homelessness: some figures show that there are fewer homeless people in most Western cities today than two decades ago. But results differ when studies include precarious and temporary forms of shelter. This statistical ambiguity also applies to concepts of resilience and fragility. What is the opposite of a fragile system? Some of our participants were not convinced that fewer disaster-related deaths and injuries constituted a reliable indicator of more resilient systems. They argued that such data fail to capture other rising forms of human suffering. Similarly, participants asked about how, and at what levels, progress should be measured. When statistics are based on samples drawn at national and global levels, they often mask the suffering and vulnerability of individual communities and social groups.

Three Significant Differences

Given this context, there are three significant differences evident in the opinions of participants. The first difference concerns the role of technology. Is technological advancement increasing our ability to mitigate risks—such as those brought on by climate change—or is it making us more vulnerable? Participants and panelists tended to agree that technological advancement has made work, construction, travel, and communication safer and easier, overall leading to a better quality of life. But some participants identified two problems with the issue of technological progress. First, progress is making us increasingly dependent on computer-based systems. Second, progress often happens at the expense of the stability of ecosystems and nature.

Another significant difference among participants regards the question of connectivity. Does the complexity, interdependence, and interconnectedness of contemporary systems produce greater resilience or fragility? Some argue that increased connectivity produces greater vulnerability because failure in any area can bring down an entire system. For instance, an Internet failure would have drastic consequences for transportation and communication systems. But according to other participants, the chances for the recovery of a system rise alongside its relative complexity and interconnectivity. Following this idea, a disaster-affected community has better chances of recovery if it already features strong social connections and communication channels. One of the blog commenters concluded that the question "is not whether connectivity is a source of resilience, but how it can be a source of resilience."

A third difference concerns the tension between progress and social justice. Most participants agree that as much as progress is real, it is not equal, and does not systematically lead, to socially-just outcomes. Lower-income individuals often lack sufficient access to transportation, communication, health services, and full employment, making them particularly vulnerable in the face of disaster. Freedom of movement and travel between countries is a privilege of the wealthy. Cities are no longer walled, but spatial exclusion still targets disadvantaged and undesirable populations. Many would probably feel conflicted about Thomas Fisher's advice of "living frugally, working locally, and traveling as little as we can," since they already do so.

Some participants fear that a narrative of progress can be mobilized to ignore urgent local problems. Scholars and practitioners must ensure this inequity does not happen. Urban progress is an important area of study, but it must be framed and nuanced by principles of social justice and environmental responsibility.

NOTES

1. Saleemul Huq and Mozaharul Alam, "Flood Management and Vulnerability of Dhaka City," in *Building Safer Cities: The Future of Disaster Risk*, ed. Alcira Kreimer, Margaret Arnold, and Anne Carlin (Washington, DC: World Bank, 2003); Mark Pelling and Ben Wisner, eds., *Disaster Risk Reduction: Cases from Urban Africa* (London: Routledge, 2012).
2. Mark Jayne, *Cities and Consumption* (London: Routledge, 2005).
3. Zhifeng Liu et al., "How Much of the World's Land Has Been Urbanized, Really? A Hierarchical Framework for Avoiding Confusion," *Landscape Ecology* 29, no. 5 (2014): 763–771; see also "Cities and Climate Change," United Nations Environment Programme, 2020, https://www.unenvironment.org/explore-topics/resource-efficiency/what-we-do/cities/cities-and-climate-change.
4. Global Facility for Disaster Reduction and Recovery (GFDRR), *A Decade of Progress in Disaster Risk Management*, 2017, https://www.gfdrr.org/sites/default/files/10%20Years%20DRM%20Development%20DRAFT.pdf.
5. Jeremy Travis and Michelle Waul, "Reflections on the Crime Decline: Lessons for the Future," paper presented at the Proceedings of the Urban Institute Crime Decline Forum, Urban Institute Justice Policy Center, Washington, DC, 2002; Joshua S. Goldstein, *Winning the War*

on War: The Decline of Armed Conflict Worldwide (New York: Penguin, 2012).

6. Theodore Levitt, "The Globalization of Markets," in *Readings in International Business: A Decision Approach*, ed. Robert Z. Aliber and Reid W. Click. (Cambridge: MIT Press, 1993).
7. Steven Pinker, *The Better Angels of Our Nature: The Decline of Violence in History and Its Causes* (New York: Penguin, 2011).
8. Intergovernmental Panel on Climate Change (IPCC), *Climate Change 2014: Impacts, Adaptation and Vulnerability*. Working Group II Contribution to the Fourth Assessment Report of the Intergovernmental Panel on Climate Change (Cambridge: Cambridge University Press, 2022).
9. C. S. Holling, "Engineering Resilience Versus Ecological Resilience," in *Engineering Within Ecological Constraints*, ed. P. C. Schulze (Washington, DC: National Academy Press, 1996), 31–44.
10. Eric Klinenberg, *Heat Wave: A Social Autopsy of Disaster in Chicago* (Chicago: University of Chicago Press, 2002).
11. Lance H. Gunderson and C. S. Holling, eds., *Panarchy: Understanding Transformations in Human and Natural Systems* (Washington, DC: Island Press, 2002).
12. Jared Diamond, *Lessons from Environmental Collapses of Past Societies* (Washington, DC: National Council for Science and the Environment Washington, 2004).
13. See Luis W. Alvarez et al., "Extraterrestrial Cause for the Cretaceous–Tertiary Extinction," *Science* 208, no. 4448 (1980): 1095–1108; Shelby Lyons et al., "Organic Matter from the Chicxulub Crater Exacerbated the K–Pg Impact Winter," *Proceedings of the National Academy of Sciences* 117, no. 41 (2020): 25327–25334; David A. Kring, "The Chicxulub Impact Event and Its Environmental Consequences at the Cretaceous–Tertiary Boundary," *Palaeogeography, Palaeoclimatology, Palaeoecology* 255, no. 1–2 (2007): 4–21; Peter Schulte et al., "The Chicxulub Asteroid Impact and Mass Extinction at the Cretaceous-Paleogene Boundary," *Science* 327, no. 5970 (2010): 1214–1218; William F. Bottke, David Vokrouhlický, and David Nesvorný, "An Asteroid Breakup 160 Myr Ago as the Probable Source of the K/T Impactor," *Nature* 449, no. 7158 (2007): 48–53; Frank T. Kyte, "A Meteorite from the Cretaceous/Tertiary Boundary," *Nature* 396, no. 6708 (1998): 237–239.

14. Jared Diamond, *Collapse: How Societies Choose to Succeed or Fail* (New York: Penguin, 2005).
15. Garrett Hardin, "The Tragedy of the Commons," *Science* 162, no. 3859 (1968): 1243–1248.
16. Myron Weiner, "Security, Stability, and International Migration," *International Security* 17, no. 3 (1992): 91–126; International Organization for Migration (IOM), *World Migration Report 2020* (Geneva: IOM, 2019).

2

ON POWER IMBALANCES

Is Disaster-Related Research and Practice in the Global South Unfavorably Guided by Northern Ideas?

WITH J. C. GAILLARD AND CARMEN MENDOZA-ARROYO

Low-income countries—sometimes collectively termed the Global South—suffer the most from disasters and the effects of global warming. Yet, as many experts note, ideas developed by decision-makers and intellectuals from rich countries—the Global North—dominate research and policy in disaster reduction and response. Many such experts believe that scholars of disaster studies adopt, whether intentionally or not, a form of neocolonialism in their academic work, with numerous unfortunate consequences for theory and practice.[1] A central element is that much scholarship mirrors uneven power relationships between the Global North and South, between "the West and the Rest." Local knowledge is often overlooked, and research and methods, from the stance of critics, are overinfluenced by Western or Northern concepts.[2] In the countries of the Global South, researchers and decision-makers filter local conditions through these international lenses. This leads to policies and projects that rarely fit the needs and expectations of poor communities in the South. Moreover, local and indigenous solutions are often scorned and replaced with imported concepts and paradigms like resilience, sustainability, adaptation, and informality. Critics claim that imported concepts are at best useless, and at worst dangerous, and argue for an intellectual and moral reform that repositions research away from Western ideas.[3]

Not all experts agree. Some find that focusing on a researcher's birthplace, nationality, or place of residence is flawed.[4] For these experts, proximity to people, places, or situations under analysis does not guarantee the quality of scholarship or the pertinence of resulting ideas. The value of one's scholarship depends on the rigor of one's research methods and the depth of understanding of the context under study. These scholars also contest allegations of Western academic colonialism and note that, at present, China is one of the most prolific centers in disaster studies. Moreover, they argue that academic marginalization does not exist solely in relations between the Global North and South; power imbalances also occur between researchers in the same country and are based on unequal center-periphery relations rather than distinctions in national wealth. Some experts argue that the distance between Western scholars and the contexts they investigate enables them to provide insightful readings of local situations through original approaches and from unforeseen perspectives. Like local academics, they may use their knowledge and prominence to reveal social injustices and fight for the rights of vulnerable populations in the South. In their view, the competitive model undergirding academia, in both low-income and wealthier countries, rather than "academic colonialism," is the real enemy.

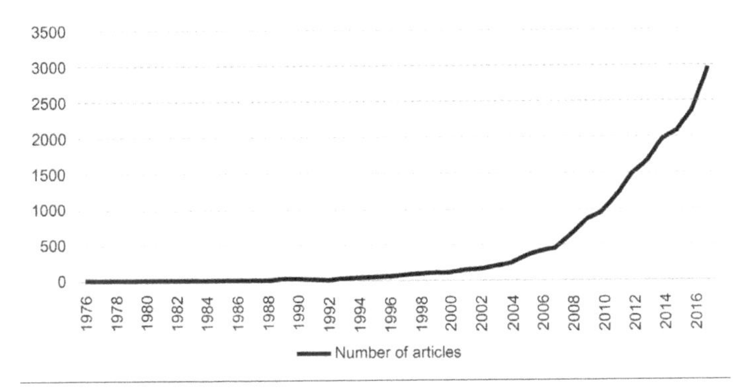

FIGURE 2.1 Yearly number of social science journal articles referring to "disaster" and "vulnerability" between 1976 and 2016 (according to Scopus).

J. C. GAILLARD: DISASTER-RELATED RESEARCH AND PRACTICE IN THE GLOBAL SOUTH IS UNFAVORABLY GUIDED BY NORTHERN IDEAS

In 1976, when Phil O'Keefe and colleagues published "Taking the Naturalness Out of Natural Disasters," only two social science journal articles had referred to "disaster" and "vulnerability" to frame their arguments.[5] In contrast, in 2018, the Scopus database of academic papers found those same concepts used in 2,940 articles (figure 2.1). This sharp increase reflects how popular and widespread these concepts have become in disaster scholarship.

Disaster and *vulnerability* are terms that etymologically originate in Latin. They do not translate well into many non-Latin languages. The hegemonic use of these concepts in disaster studies reflects the dominance of a particular understanding of and approach to disasters over other ways of framing harm, hardship, and suffering. It also reflects the dominance of English in publishing on disasters and the translation of concepts beyond their original contexts. This debate therefore centers on ontological and epistemological issues and on power relations.

The spread of Latin-based disaster-related concepts such as disaster and vulnerability, in addition to resilience, adaptation, and capacity, among others, mirrors the hegemonic influence of Western ontologies and epistemologies within disaster studies. While these concepts are diverse, they are rooted in a common legacy inherited from the Enlightenment. At stake is the widespread adoption of these concepts outside of their original context, a practice that reflects and perpetuates both colonial and neocolonial histories. To use Edward Said's word, this approach promulgates an "Orientalist" view of disasters.[6]

Large disasters are occurring with greater frequency in regions where people share diverse understandings of the world. Many of these understandings have been shaped by unique intellectual traditions that build upon epistemologies—theories of knowledge—that are as rigorous and consequential as those inherited from the Enlightenment. However, they are often overwhelmed by outside perceptions of disasters, as though, to echo Frantz Fanon, the adoption of Western approaches signaled elevated status and more rigorous scholarly values.[7]

The hegemony of Western ontologies and epistemologies in disaster studies has underpinned a normative agenda for disaster risk reduction. It has sustained decades of international policies that fostered the transfer of experience and resources from the West to the rest of the world, thus further skewing power relations. On the ground, these policies materialize in standardized practices that filter disaster risk assessment and reduction initiatives through the lens of concepts that cannot adequately capture people's experience of what researchers commonly call "disasters."

An obvious example of such skewed practices is the so-called vulnerability and capacity assessment (VCA) toolboxes. These toolboxes rely on taxonomic categorizations of individual resources and identities according to age, gender, and physical abilities. These categories, often associated with quantitative and/or demographic indicators and based on preconceived ideas about lived experience, are organized as fillable boxes. In the case of gendered approaches to assessing and reducing disaster risk, the VCA toolboxes cannot appropriately address nonbinary identities. The hegemony of Western ontologies and epistemologies ultimately contradicts the essence of the paradigm advanced in "Taking the Naturalness out of Natural Disasters" and its associated commitment to the individual's participation in reducing disaster risk. We nonetheless often embrace this paradigm when using concepts such as disaster and vulnerability to frame scholarly studies, an intriguing incongruity that may minimize our adequate assessment of local and indigenous perspectives.

CARMEN MENDOZA-ARROYO: DISASTER-RELATED RESEARCH AND PRACTICE IN THE GLOBAL SOUTH IS NOT UNFAVORABLY GUIDED BY NORTHERN IDEAS

The challenge for professionals—national and local, not only international—is how to engage with the majority of construction that happens as non-engineered structures in informal, unplanned settlements. This is where the major risks and vulnerabilities are.

—Graham Saunders[8]

I would like to begin with this quote by the late humanitarian architect Graham Saunders. He emphasizes that most reconstruction is carried out by families and local builders. It is their capacity that needs to be developed to achieve safer buildings. In other words, the first and most important requirement in dealing with disaster recovery is that local communities in affected regions be seen as active collaborators, rather than helpless beneficiaries of aid relief. As such, our work must create opportunities for capacity building so that local professionals can assume complex processes; this must be a priority addressed before disasters occur. Likewise, one of the major challenges of our profession is the cocreation of knowledge, knowledge sharing, and its translation into local building codes in disaster-prone countries so as to introduce better and safer construction processes.

The reason disaster studies and practices are questioned as overinfluenced by Western concepts and dominated by ideas developed in rich countries is that we very often forget that architecture and urbanism must promote the uniqueness of a place; in order to accomplish this, we must promote its culture.[9] From this perspective, working with the causes, history, and cultures in which we intervene can facilitate local resilience. When a sense of place is reinforced in efforts to reduce disaster risk or recover from disaster, a possibility for people's autonomy emerges, despite the implementation of technical Western concepts.[10] In other words, strengthening capacities in disaster recovery can break long-term legacies of dependency, as user-built reconstruction promotes safer, better buildings and enhances community development.[11]

As professionals and academics, we must understand the limitations of our expertise and knowledge, thereby avoiding any bolstering of relationships of dominance.

Furthermore, it is vital to acknowledge that local communities need active and professional support in devising their recovery and building in a safe and efficient manner. Today, design professionals must accept that the disaster field is merging with other areas, such as climate change adaptation, displacement, and migration. This shift frames a complex working field and reinforces an urgency for an adaptive approach to design that draws social, technical, and community concerns into urban crisis responses. Such a complex approach demands a multifaceted, multiscale, and comprehensive view of the built and natural environment, in addition to more resources and broad expertise.

Perception is a constant factor that amplifies or mitigates risk. Risk perception closely connects to culture and ideology. A belief that work is being done to serve people has a positive influence, while work is often perceived as bad or corrupt if it is completed in the service of vested interests. As such, a foreign perspective provides distance and attention to the root causes of risk, thereby introducing a comparative view and broader scope to its reduction and management.

In summary, it is important to recognize and increase the capacity of local groups and professionals through horizontal learning experiences that strengthen ties among local organizations, institutional collaborators, and global professionals.

J. C. GAILLARD'S REBUTTAL REMARKS

The debate has partly shifted from considerations of ontologies, epistemologies, and power relations to those on the positionality of researchers. The two are inextricably linked, as researchers produce studies that perpetuate hegemonic ideas in their approaches to disaster studies, which then filter down

PARTICIPANT'S COMMENTS

Ekatherina Zhukova, Senior Lecturer, Department of Political Science, Lund University, Sweden

To address this question, it might be more beneficial to discuss privilege instead of distinctions between North and South. Both Northerners and Southerners can have privileges in a sense of social capital (language, finance, networks). The leading question might then be reframed along the lines of: "Is disaster-related research and practice unfavorably guided by a researcher's or practitioner's privilege?" Or, according to a postcolonial approach, the question might ask who can speak and act, for whom, how, and why? And who can't speak and act? The subaltern exists in both the Global North and South.

Since I moved to the Northern part of the world, I have been puzzled about where the former Soviet space belongs? Is it Northern, Southern, "poor" Northern, or neither? It has seemingly been left out of the debates on the North-South dichotomy. I have also noted the recurrence of terms such as "First World," "Second World," and "Third World" to define global development and, relatedly, terms such as "developed," "developing," and "underdeveloped." I have been researching the Chernobyl nuclear disaster, and many authors refer to the affected space as "post-communist," "post-socialist," "countries in transition," and so on.

How can we understand disasters in their own right, outside preestablished categories? If we establish disaster-related categories not linked to existing indicators such as income, human development, governance, innovation, or geography, can we be sure that we will not create new power relations? Every category presupposes an exclusion.

to the policies and practices geared toward reducing the risk of disaster.

Imbalanced power relations among researchers lead to the decontextualization of concepts such as disaster and vulnerability from more specific regional, ontological, and epistemological frameworks. As a result, comprehension of disasters may misrepresent the lived experiences of those who directly confront so-called hazards, as demonstrated by Mihir Bhatt in his seminal essay on vulnerability.[12]

A recent survey of the papers published in the journal *Disasters* makes this scholarly imbalance clear: 84 percent of authors are associated with institutions located in countries of the Organisation for Economic Co-operation and Development (OECD) (an imperfect proxy for the West), which are less affected by large disasters.[13] The imbalance is particularly evident when looking at high-profile disasters such as the 2010 Haiti earthquake or the 2015 Nepal earthquake (see figure 2.2). Research on disasters is therefore largely dominated by researchers foreign to their geographical area of study.

However, a researcher's institutional affiliation or residence need not be based in places where disasters occur; a local perspective is not required to conduct research on the topic. Some Haitian and Nepalese researchers based in Europe, North America, and Australasia have been conducting studies in their native land. Yet in most cases, they rely upon Western ontologies and epistemologies centered on concepts such as disaster and vulnerability, as some local researchers affected by these disasters have done as well. This tendency toward hegemonic approaches to disaster research reflects Jacques Rancière's concept of intellectual submission, in this case by prioritizing ontologies and epistemologies inherited from the Enlightenment.[14]

Unequal power relations between researchers are sustained by intersecting interests. Principal investigators and their allies in

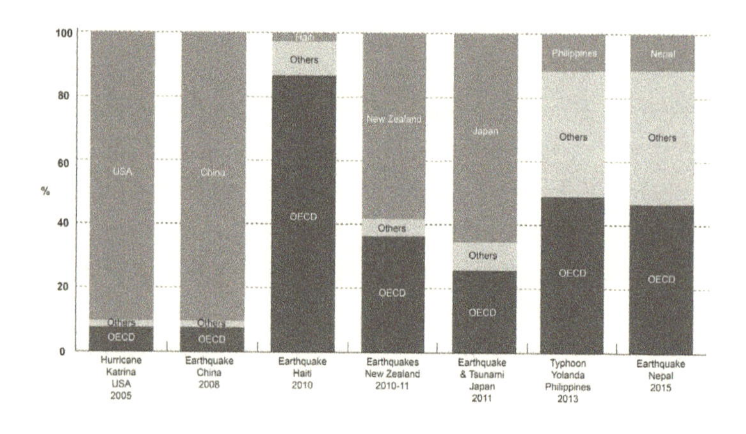

FIGURE 2.2 Country of affiliation of lead authors of papers published on high-profile disasters between 2005 and 2015 (data from Scopus; reprinted with the permission of John Wiley and Sons).

Western countries, where most research funding is available (with the exception of China), may give up some of their funding to co-investigators based in institutions near places affected by disasters. The latter gain opportunities to develop collaborations, access expensive equipment, and publish in international journals, thus advancing their careers.

Both the intellectual submission and the coincidence of interests that define research collaboration are crucial to understanding the hegemony, in Antonio Gramsci's terms, of ontologies and epistemologies in disaster studies inherited from the Enlightenment and their influence on disaster risk reduction policies and practices.[15] Such unequal power relations between researchers similarly define scholarly imbalances within individual countries. Scholars affiliated with universities located in national capitals and large cities often have more access to resources and broad academic networks; they are more likely to adopt Western ideas and approaches, as well as exert control over institutions and scholars at the periphery. This reflects the Westernization of an increasing "competitivity" within academia, as Alexander rightly emphasized, as well as the wide-reaching colonial and neoliberal heritage of the Global North.[16]

PARTICIPANT'S COMMENTS

Vanicka Arora, PhD candidate, Institute for Culture and Society, Western Sydney University, Australia

I certainly agree that the world is far more entangled than suggested by the standard dichotomy of North and South. I am an Indian based in Australia, and my research is situated in Nepal. Does this make me Northern or Southern? By this definition, the funding I receive, my supervisors, and my institution are Northern, but my background and cultural identity are Southern. And certainly, being Indian has not made me any more qualified to look at Nepal, nor the precise town on which my research focuses. The post-disaster research landscape in Nepal, which is its own unique ecosystem, currently comprises a dizzying array of countries, networks, and funding agendas, to the extent that I need to maintain a database of the different research agendas in play (in a table that is often out of date). So many partnerships operate among multiple universities, NGOs, and "local" institutions, that it is unproductive to evaluate them using a North/South paradigm. A more productive line of inquiry might involve looking into the actors responsible for establishing research agendas. Is research co-constituted between the researcher and the researched? To what extent does funding influence outcomes?

Nation-state relations can be and are resisted through internal politics, and vice versa. For instance, while India has invested heavily in post-disaster recovery in other countries, it resisted foreign aid offered to Kerala following the 2021 floods. Government officials responded prematurely to unverified claims, a decision greatly influenced by national politics and international appearance. Similarly, the city of Bhaktapur in Nepal famously turned down German aid for reconstruction, thereby illustrating how local politics can bypass the nation to respond directly to international politics. This is not to diminish the continuing influence of past and present colonial regimes, but to flag that the "poor nation" is also a construct that can be resisted in different ways.

Disaster scholarship has been shaped incongruously. Many seem to embrace a paradigm advanced forty years ago to challenge the then-dominant and hazard-driven understanding of disasters.[17] The authority of local researchers and communities, outside of the academic silo, defined this paradigm. As such, this paradigm encouraged studies grounded in local ontologies and epistemologies. In practice, however, it seems evident that such an approach has not yet flourished.

CARMEN MENDOZA-ARROYO'S REBUTTAL REMARKS

I think we can all agree, based on the insightful contributions of the participants and J. C. Gaillard, that certain policies, practices, handbooks, and other tools have encouraged the application of standards and experience from the West to the rest of the world as rigid guides to recovery, excluding considerations of local adaptive capacity in disaster recovery. However, that is only part of the issue at hand.

I was born in Bolivia and lived my formative years in various Latin American countries. These Latin American roots and culture are central to my personal and professional identity. I developed research tools and methodologies in both Latin America and Barcelona. My background enables me to develop my research on two continents, and my approaches to be nurtured in both worlds. Therefore, I believe the present debate should not exclusively circumscribe the geographic origin of policies or concepts, but should also consider the inadequacy of blanketly adopting standardized assessments and recovery frameworks, given the diversity of governments, cultures, development, and access to resources of regions pursuing recovery.

We agree that effective and culturally specific recovery depends on localized, adapted frameworks of response instead of global approaches. However, regarding the question of "universal shelter standards versus national standards," Peter Walker introduces a compelling argument regarding Sphere Standards.[18] He discusses how, despite their implementation without consultation with national governments and with subsequent disregard of local cultural and economic factors, their practical value resides in establishing minimum standards in shelter and housing. Had the Sphere Standard not been implemented, unregulated substandard work might have resulted in hardship for survivors.

Communities dealing with or anticipating disasters prioritize post-disaster recovery but do not necessarily consider disaster risk reduction. Knowledge, innovation, and education are necessary to build a culture of safety and resilience and to strengthen disaster preparedness. Furthermore, "accountability and trust" rather than control should motivate international donors, who frequently drive recovery decisions and implementation. The "trust-control dilemma" presented by Ian Davis and David Alexander, based on Charles Handy's earlier work, affirms that international agencies' agendas and frameworks reflect a desire to control at the expense of trust.[19] From this perspective, the real issue concerns donors' perception of the level of governance and trust in national and local governments, which influences and shapes their recovery guidelines.

Finally, I would like to underline that post-disaster reconstruction is a social process as much as a technical one, and recovery must be professionalized so that the technical response is suitable. The impact of "natural" disasters is always local, and the response should follow suit.

J. C. GAILLARD'S CLOSING REMARKS

The momentum around this debate shows that the time is ripe for a critical reflection on how disaster studies inform policies and practices devised to reduce disaster risk. More than five hundred people have expressed their agreement with the recent disaster studies manifesto "Power, Prestige & Forgotten Values," which calls for such a reflection. The document, authored by a team of twenty-four scholars and practitioners from around the world, suggests three key areas of reflection.

First, research on disaster should focus on whatever is needed and deemed relevant at a local scale. Enrico Quarantelli and Russell Dynes have long held that disasters are local issues.[20] Of course, disasters result from and reflect broader processes that underpin their root causes—processes that have been extensively researched over the past four decades. However, understanding what disasters are, their impacts, and how people deal with them most often requires local grounding.

Second, to frame the most relevant perspective(s) on disaster studies, local grounding demands research based on local ontologies and epistemologies. We should therefore be critical of hegemonic concepts and methodologies inherited from the Enlightenment. As Rohit Jigyasu wrote, "Our understanding of disaster needs to be turned inside out and not the other way around, as it tends to become, thanks to the 'expert' notions of what is a disaster."[21] It is essential to engage in the "revolution in thinking about disasters" called for by Ben Wisner and colleagues in 1976.[22]

Third, we should ultimately support local researchers, who best understand local contexts and are likely to be familiar with local ontologies and epistemologies. They should take the lead in studying local disasters. Local scholars should become principal investigators of projects and lead authors of research output, and they should likewise favor local outlets and venues for research dissemination to affect local policies and practices. Local researchers should also prioritize participatory approaches involving local people to answer Robert Chambers's call to put the last first.[23]

This alternative approach to researching disaster calls for engagement with subaltern studies, as pioneered by Ranajit Guha and Gayatri Spivak in South Asia.[24] Many fields of study, such as indigenous studies, history, and psychology, have charted valuable approaches that integrate subaltern perspectives.[25] For disaster studies to follow a similar direction, the discipline must shift away from its intellectual submission to ideas inherited from the European Enlightenment. This shift first requires consciousness, in Freire's terms, then emancipation, in Rancière's terms, among subaltern groups, here applying to researchers whose local understandings of the world differ from Western perspectives.[26]

Importantly, this approach does not preclude hybridity and collaborations with outside researchers, especially when they have built trust with local peers. As much as possible, the latter should lead and the former support, recognizing that, in some contexts, the voices of local researchers may be filtered by state power. Similarly, Enlightenment concepts and frameworks should not be ditched outright. They are still useful to uncover global processes associated with the root causes of disasters and related normative disaster risk reduction policies. Furthermore, they often coexist with local understandings of harm, hardship, and suffering with hybrid cultural environments.[27]

In summary, disaster studies should not become an exclusive area of research. It should concern itself with enhancing diversity, hybridity, dialogue, and collaboration to ensure that the most appropriate approaches and ideas are employed to understand disasters and inform policies and practices geared toward reducing risk in local and global contexts.

CARMEN MENDOZA-ARROYO'S CLOSING REMARKS

This debate's original question shifted to tie into larger reflections that go beyond geographic perspectives, namely, to what extent do disaster-related research and practice emerge from unsubstantiated ideas and unequal power relations on national and international scales? In what ways do such dynamics unfavorably guide research and practice? Do they indeed overlook local knowledge, cultures, and practices due to factors like corruption and inequality?

I have argued that local populations are key to the sustainable improvement of their built environment, since their lived experience informs capacities, knowledge, resilience, and ultimately reconstruction. Although most reports, guidelines, or manuals previously advocated for standardization, the discussion in recent international NGOs' reports centers on the failure of culturally foreign guidelines; such reports instead advocate for context-adapted, "situated solutions," proclaiming that "assessment is the foundation for appropriate response" for housing and urban areas.[28] Clearly, adjustments to existing local, social, technical, and financial organizational capacities, rather than the provision of appropriate technical responses, are the most important factors for recovery. I would like to emphasize one factor that advances noninclusive research and practice: the lack of gender-specific policies and practices that have otherwise been dominated by ideas developed by male practitioners and decision-makers. It is imperative to explore existing gender roles in local communities and address the experiences of women to successfully design equitable long-term strategies and recovery plans.

John Mitchell's Four C's model offers an interesting perspective on the ways foreign aid exercises its power in post-disaster recovery: comprehensive, where there is no local capacity, as with the earthquake in Haiti; constrained, where operations are limited, as with the Syrian crisis; collaborative, where agencies "fill in the gaps" that the government cannot cover, as with the Pakistan floods of 2010–2011; and consultative, which provides support to strong local governments, as with the 2010 Chilean earthquake.[29] Roughly 50 percent of foreign aid spent to date has focused on a comprehensive solution, while less than 5 percent has been spent on consultative solutions.[30] An appropriate response would involve adopting only collaborative and consultative approaches comprised of providing support, building skills, and enhancing existing resources; however, this response is not always feasible.

PARTICIPANT'S COMMENTS

Duvan Hernán López, PhD student, Sustainability, Polytechnic University of Catalonia, Spain

North/South binaries should be approached not from a nationalistic perspective but with reference to the global concentration of power, wealth, control, and sovereignty and the global distribution of social and environmental costs and impacts of development. The hegemony of the American dollar as the central currency for trade is one among many possible examples of this approach. The commercial transactions, consumption choices, and chains of value of wealthy individuals in India, South America, Russia, or China always transfer their benefit to a central power, in this case the U.S. Federal Reserve.

Centrality of power constrains autonomy of power, which conditions modes of operability. I cannot imagine any post-disaster intervention escaping this logic of accumulation of value, at least without an explicit determination to redistribute and democratize the means of production through recovery processes. The danger of biased ideas in disaster research and practice is not necessarily caused by a foreign origin. I acknowledge the importance of valuable insights from Northern intellectuals and practitioners with positive incidence in disaster management. Disaster could probably provide opportunities to experiment and develop other types of economic relationships. Critiquing Northern interests and agendas, rather than ideas, is key—I am less concerned with the origin of ideas than the origin of the interests they serve.

I began this debate by maintaining that culturally based approaches embodying local traditions and practices must be implemented and better understood in terms of risk reduction. The World Disasters Report by the International Federation of Red Cross and Red Crescent Societies (IFRC) also identified the important role of culture and its centrality in understanding and managing urban disasters.[31] However, as demonstrated by the example of populations living on the slopes of La Paz, Bolivia, the defense of local knowledge and experience is not enough. Despite the organization of the community on slopes to avoid risk-prone land, these settlements have "overwhelmed urban planning and as a consequence, disasters have increased, with loss of life, houses, and livelihoods."[32] Therefore, in countries where resources are limited and where policy, planning, and delivery mechanisms are weak, local communities and their cultures must be provided with the resources to make better choices, and their advocacy for protective and inclusive laws and planning codes must be supported. Finally, reconstruction projects that understand situated social dynamics must be prioritized as the key ethical approach to design in the humanitarian field.

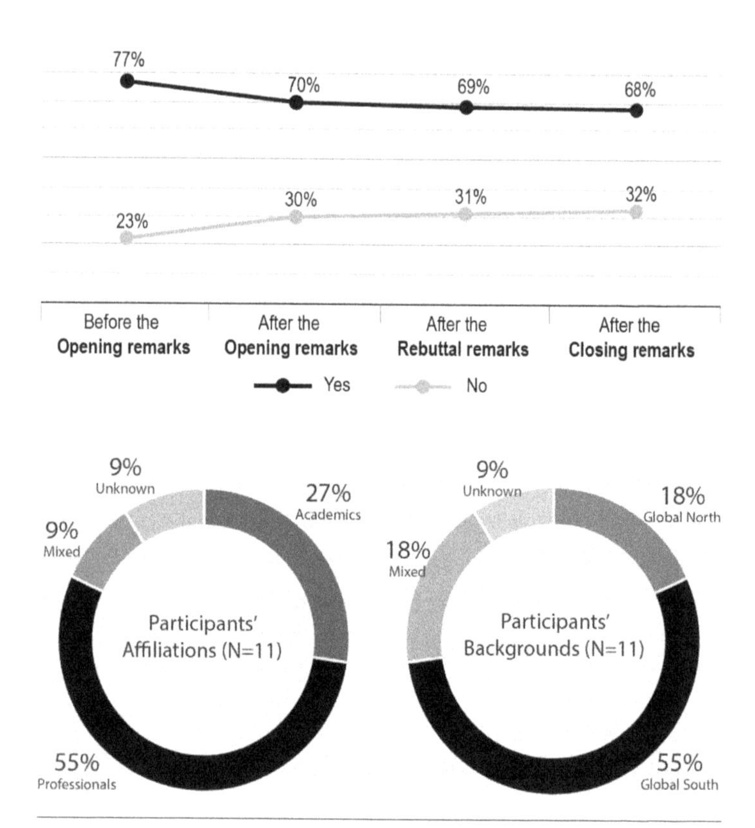

FIGURE 2.3 (*Top*) Results of votes at each stage of the debate. (*Bottom*) Distribution of affiliations and backgrounds among participants in the discussion.

THE MODERATOR'S CLOSING REMARKS: A CORE AND A PERIPHERY IN DISASTER-RELATED RESEARCH AND PRACTICE

This debate tackled how research and practice in the disaster field replicates power imbalances among and within nations. Many saw radical change as needed to balance power around disaster research and practice.

In this online debate, about 70 percent of participants consistently argued that a form of colonialism and Northern intellectual hegemony characterizes disaster research and practice. Both sides of the debate agree that change is urgently needed. Disaster-related research and practice should focus on local needs, expectations, knowledge, and resources, with an emphasis on preserving traditions, values, and a sense of place. Importantly, local researchers and practitioners familiar with local contexts, ontologies, and epistemologies must be trusted and supported.

But panelists' and participants' opinions on whether disaster studies and practices are unfavorably guided by the North raise additional questions and require certain distinctions. Two main challenges were exposed through this debate.

The first challenge concerns sources of ideas in disaster studies. Whereas it is relatively easy to identify the origins of military and political colonialism and imperialism, it is harder to determine the origin of disaster-related ideas, epistemologies, and methodologies. Duvan López, a participant, rightly reminded us that intellectual hegemony is generated from "power imbalances that occur between nations; but also, from inside countries, neighborhoods, and households." Participants also asserted that some influences are rooted in power relations that transcend the classic dichotomies of North and South, the West and the Rest. In this vein, we explored South-North and South-South influences, pointing to examples where the transfer of ideas occurs in more than one direction. Some participants argued that unequal power relations exist between researchers and practitioners, regardless of their countries of residence and professional affiliations. Power inequalities are based on several elements, such as core-periphery political relationships, connections to professional networks, funding, visibility, and access to social media.

A consensus emerged that some form of complicity can, at times, exist between "external" researchers and practitioners in the disaster field and their political and economic agendas. Are these influences a form of systematic international colonialism, or just ordinary misunderstanding and corruption? Answering this question also proved difficult. Disaster capitalism, for instance, is imposed by foreign leaders and corporations, but it is also imposed by local politicians and companies that, attracted by the lure of financial profit, often disregard the root causes of vulnerability. Other forms of domination based on local political partisanship, for instance, can also neglect the affected populations' needs and aspirations, local practices, cultural beliefs, meanings, and values.

The second challenge concerns the classification of researchers and practitioners. Whereas it is easy to agree on the need for new methods and frameworks based on insiders' knowledge and practices, it is more difficult to determine who should be considered an "insider" or "outsider." Several problems arise when the discussion moves from the existence of Northern/Western ideas to who, exactly, embodies those ideas. Most of us find it easy to label ideas as "Northern," "external," or "alien," but we sometimes find it more difficult to classify researchers and practitioners under local/external dichotomies. Can researchers and studies be classified according to geographic locations? Can we ignore the uniqueness of each researcher and the complexity of their identity, background, ethics, values, and methods when confronting scientific imbalances?

Are we mistakenly labeling individuals when we claim that there is a need to support local researchers and let them lead the study of "their" local disasters? Are we, as Vanicka Arora, a participant, argued, proceeding to an "othering" of sorts by presenting the "local" researcher as free from global influence—in essence, as a nonpolitical actor? Vanicka also insightfully pointed out that the common assumption that "the local exists as independent from the global" is wrong. It is also mistaken to demonize the North and assume that, in opposition to it, "there exists a cohesive 'local,' which is less prone to inequality, corruption, and power dynamics."

Despite these challenges, most participants agreed that intellectual elitism in research studies and practices is real and dangerous. But whereas statistics show that most disaster studies are products of the North, the dominance of key concepts and paradigms is not only the result of North-South relations. Like any other form of social injustice, power imbalances in disaster-related research and practice are rooted in differences that exist not only among nations but also within countries, cities, communities, and organizations. A paradigm shift is needed in disaster studies and practice; our approaches should originate from the knowledge, understanding, and respectful consideration of the context under study.

NOTES

1. Philip G. Altbach, "Globalisation and the University: Myths and Realities in an Unequal World," *Tertiary Education & Management* 10, no. 1 (2004): 3–25.
2. J. C. Gaillard, "Disaster Studies Inside Out," *Disasters* 43 (2019): S7–S17; Danielle Zoe Rivera, "Disaster Colonialism: A Commentary on Disasters Beyond Singular Events to Structural Violence," *International Journal of Urban and Regional Research* 46, no. 1 (2020): 126–135.
3. Ernesto Aragon-Duran et al., "The Language of Risk and the Risk of Language: Mismatches in Risk Response in Cuban Coastal Villages," *International Journal of Disaster Risk Reduction* (2020): 101712.
4. David E. Alexander, "Disaster Studies Outside In," *Disaster Planning and Emergency Management*, January 5, 2019, http://emergency-planning.blogspot.com/2019/01/; Gonzalo Lizarralde, et al., "We Said, They Said: The Politics of Conceptual Frameworks in Disasters and

Climate Change in Colombia and Latin America," *Disaster Prevention and Management* 29, no. 6 (2020).

5. Phil O'Keefe, Ken Westgate, and Ben Wisner, "Taking the Naturalness Out of Natural Disasters," *Nature* 260 (1976): 566–567.
6. Edward Said, "Introduction to Orientalism," in *Media Studies: A Reader*, ed. Sue Thornham, Caroline Bassett, and Paul Marris (New York: New York University Press, 1978), 111–123.
7. Hussein Abdilahi Bulhan, *Frantz Fanon and the Psychology of Oppression* (New York: Springer Science & Business Media, 1985).
8. RMIT et al., *Creation and Catastrophe: Symposium Report*, April 7, 2016, https://harbureau.org/downloads/1-Creation-and-Catastrophe-Final-Report-online.pdf.
9. Lizzie Yarina, "Your Sea Wall Won't Save You: Negotiating Rhetorics and Imaginaries of Climate Resilience," *Places Journal*, March 2018.
10. J. C. Gaillard, "Disaster Studies Inside Out," *Disasters* 43 (2019): S7–S17.
11. Ian Davis, "Reducing Disaster Risks 1980–2010: Some Reflections and Speculations," *Environmental Hazards* 10, no. 1 (2011): 80–92.
12. John Twigg and Mihir R. Bhatt, *Understanding Vulnerability: South Asian Perspectives* (Intermediate Technology Publications, 1998).
13. Gaillard, "Disaster Studies Inside Out."
14. Jacques Rancière, *Le maître ignorant: Cinq leçons sur l'émancipation intellectuelle* (Paris: Fayard, 2014).
15. Antonio Gramsci, *Prison Notebooks*, vol. 1 (New York: Columbia University Press, 1992).
16. Alexander, "Disaster Studies Outside In."
17. Gaillard, "Disaster Studies Inside Out."
18. Peter Walker, "Developing Minimum Performance Standards in Humanitarian Relief: Why Bother?" *Humanitarian Practice Network* 6 (1996); Sphere Association, *The Sphere Handbook: Humanitarian Charter and Minimum Standards in Humanitarian Response*, 4th ed. (Geneva: Sphere Association, 2018).
19. Ian Davis and David Alexander, *Recovery from Disaster* (London: Routledge, 2016); Charles Handy, "Trust and the Virtual Organization," *Long Range Planning* 28, no. 4 (1995): 126–132.
20. Enrico L. Quarantelli and Russell R. Dynes, "Response to Social Crisis and Disaster," *Annual Review of Sociology* 3 (1977): 23–49.

21. Rohit Jigyasu, "Disaster: A 'Reality' or Construct? Perspective from the 'East,' " in *What Is a Disaster? New Answers to Old Questions*, ed. Ronald W. Perry and E. L. Quarantelli (Bloomington, IN: Xlibris, 2005), 49–59.
22. Ben Wisner, Phil O'Keefe, and Ken Westgate, "Taking the Naturalness Out of Natural Disaster," *Nature* 260, no. 5552 (1976): 566–567.
23. Robert Chambers, *Rural Development: Putting the Last First* (London: Routledge, 2014).
24. Guha Ranajit and Gayatri Chakravorty Spivak, eds. *Selected Subaltern Studies* (Oxford: Oxford University Press, 1988).
25. Rogelia Pe-Pua and Elizabeth A Protacio-Marcelino, "Sikolohiyang Pilipino (Filipino Psychology): A Legacy of Virgilio G. Enriquez," *Asian Journal of Social Psychology* 3, no. 1 (2000): 49–71; Linda Tuhiwai Smith, *Decolonizing Methodologies: Research and Indigenous Peoples* (London: Zed Books, 2013).
26. Paulo Freire, *Pedagogy of the Oppressed*, 50th Anniversary ed. (London: Bloomsbury, 2018); Rancière, *Le maître ignorant*.
27. J. C. Gaillard, "The Tout-Monde of Disaster Studies," *Jàmbá-Journal of Disaster Risk Studies* 15, no. 1 (2023): 1385.
28. David Sanderson and Ben Ramalingam, *Nepal Earthquake Response: Lessons for Operational Agencies* (London: ALNAP/ODI, 2015), 15, https://www.medbox.org/preview/5550738c-2dc0-4de9-8692-05431fcc7b89/doc.pdf; GFDRR, *Towards a Resilient Future: Annual Report 2012*, January 15, 2013, https://www.gfdrr.org/en/publication/annual-report-2012; Doris Carrion, *Syrian Refugees in Jordan: Confronting Difficult Truths* (London: Chatham House, 2015).
29. John Mitchell, "From Best Practice to Best Fit," Montreux XIII, December 2014; David Sanderson, Jerold S Kayden, and Julia Leis, eds. *Urban Disaster Resilience: New Dimensions from International Practice in the Built Environment* (London: Routledge, 2016).
30. Mitchell, "From Best Practice to Best Fit."
31. IFRC, *World Disasters Report 2020: Tackling the Humanitarian Impacts of the Climate Crisis Together* (Geneva: International Federation of Red Cross and Red Crescent Societies, 2020).
32. Terry Cannon and Lisa Schipper, eds., *World Disasters Report 2014: Focus on Culture and Risk* (Geneva: International Federation of Red Cross and Red Crescent Societies, 2014), 71.

3

ON RESILIENCE

Is the Concept of Resilience Useful in the Fields of Disaster Risk Reduction and the Built Environment, or Is It Just Another Abused and Malleable Buzzword?

WITH DANIEL ALDRICH AND JONATHAN JOSEPH

The theory of resilience recognizes the inherent capacities of social systems to withstand, recover from, and adapt to adverse impacts.[1] Scholars who uphold the theory believe that resilience provides an ethical approach to understanding the fragile relationships among built, natural, and social environments. They conceive of resilience as a useful framework for understanding the unpredictability and complexities of our world and for examining notions of anticipation, adaptation, and proactive transformation in response to stressors.[2] For them, resilience emphasizes a constructive approach to research rather than one that simply identifies problems and their causes, in contrast to the theory of vulnerability, which is often regarded as its rival theory. More importantly, such scholars envision resilience as a tool for displaying the strengths and capacities of social systems and for systematically examining the long-term effects of multiple variables.[3]

By contrast, opponents of the theory of resilience raise serious doubts about its usefulness and relevance.[4] They claim that the concept is overused and thereby facilitates broad and sometimes contradictory meanings and interpretations.[5] They also note that neoliberal policy- and decision-makers have hijacked the theory to justify shifting their responsibilities toward the private sector. They also argue that increased resilience among some communities often results in increased vulnerabilities for others. Finally, critics contend that the varied definitions of resilience make it a fashionable buzzword with insufficient moral value.[6]

DANIEL ALDRICH: RESILIENCE IS A USEFUL CONCEPT IN THE FIELDS OF DISASTER RISK REDUCTION AND THE BUILT ENVIRONMENT

Resilience provides a useful lens for understanding the factors that influence a community's preparation for and response to disasters. Fieldwork, interviews with survivors, discussions with NGOs, and quantitative data show that diversity, flexibility, and human connections drive disaster risk reduction and recovery.[7] Human factors are better predictors of successful responses to and recovery from disasters than those most often considered by policymakers, such as the strength of physical infrastructure and aid received from the government or international relief agencies. Vulnerable individuals around the world—in India, Japan, and the United States, for example—have consistently looked to kin, neighbors, and nearby friends, rather than government authorities, for help in times of crisis. Disaster after disaster, we see that uniformed government personnel or police authorities are not first responders. Rather, families and neighbors are almost always first on a disaster scene—their support enhances resilience.

Following the 1995 Kobe earthquake in Japan, some two-thirds of survivors pulled from the rubble were rescued by neighbors, not police officers or self-defense force members.[8] Following Hurricane Katrina, communities with stronger internal cohesion and connections to city officials, such as the Mary Queen of Viet Nam (MQVN) neighborhood in New Orleans, fared best in rebuilding schools, stores, and homes. After the 3/11 compounded disasters in Tohoku, Japan, survivors with more connections and friends experienced less anxiety and stress than those who were more isolated. The networks of people to whom we are connected—some only tenuously, such as the friend of a friend, and others more deeply, such as a family member—serve as critical resources in times of crisis.[9] Strong social connections provide three main types of assistance that build resilience: they provide mutual aid and informal insurance; they facilitate collective action; and they help individuals make decisions about returning to and rebuilding damaged communities. As such, arguments about the overemphasis on individuals in neoliberal, Anglo-American resilience approaches neglect perspectives that center on communities. Community resilience—the networks and connections among people living in the same area—provides a means to help people collectively manage risk in a world increasingly under threat from rising temperatures, higher sea levels, and more extreme weather events.

I agree that aspects of the resilience approach that have found currency in the UK share a common cause with the neoliberal orientation of institutions such as the Department for International Development (DFID) and the World Bank. Governments regularly appropriate important concepts and ideas for their own use; no doubt they would prefer to embrace a philosophy that would allow them to do more with less. Nonetheless, such appropriation does not detract from the value of the resilience approach in the field of disaster risk reduction or humanitarian aid. Whether the neoliberal approach holds value, increases the quality of life, and moves the bottom billion out of poverty and disease is a separate debate.

JONATHAN JOSEPH: RESILIENCE IS ANOTHER ABUSED AND MALLEABLE BUZZWORD

The concept of resilience has enjoyed a huge degree of success across different spheres of policymaking. Whether it has succeeded in making a difference in these areas is a separate matter. Of course, the idea of resilience is bound to capture something of the efforts to withstand, recover from, or adapt to adverse impacts. And like associated ideas such as sustainability, well-being, and good governance, it is something we cannot possibly oppose. Rather, we must question whether anything specific to the term makes a qualitative difference in how we understand complex problems and our efforts to deal with them. Has resilience become indispensable to our knowledge and behavior?

We might consider three ways to address this question. The first concludes that resilience is no more than a fashionable buzzword with little substantial value. This can be shown by tracing the confusion surrounding the term's deployment. Different interpretations of the term appear in policy statements and strategy documents, none of which are elaborated in any significant or coherent way.[10] To overcome ambiguities in the definition of the concept, a proliferation of different types or aspects of resilience is accepted.

At another extreme, some argue that resilience represents a new way of perceiving the world, offering a fundamentally different understanding of how we operate, centered on the complexity, uncertainty, and unpredictability of the world.[11]

I instead suggest that while resilience may be more than a buzzword, it has little substance of its own and instead derives meaning from the wider discourse and practices within which it operates. It is a term, not a concept or a theory. Its explanatory power depends on its discursive environment. This does mean that the idea of resilience is always captured by laypeople's ways of understanding the world. Rather, it remains vulnerable to a neoliberal interpretation and deployment. Within this broader discourse, resilience is bound up with the modification of contemporary forms and techniques of governance that shift the burden of responsibility away from states and legal frameworks onto individuals and communities.

Understood as a means of framing problems, resilience operates in a distinct manner. It adopts a fatalistic approach to systemic crises and shocks, makes a virtue of adaptation, emphasizes the messy relationship between the social and the individual, and recalibrates our understanding of human capacities. In this respect, resilience represents a shift away from a classical liberal framework of protection and intervention; instead, it affirms a subjective capacity for learning and self-awareness as part of adaptation. Given the high costs of intervention, such an approach also makes sense in an age of austerity. The resilience approach is realistic and pragmatic in both a political and economic sense. Certainly, resilience might have more wide-ranging potential, but making it useful in the sense discussed here requires a concerted political struggle.

PARTICIPANT'S COMMENTS

Kristen MacAskill, Assistant Professor, Department of Engineering, University of Cambridge

Differences in the interpretation of what resilience means serve to question our assumptions and perspectives—I consider this a good thing. The debate certainly provokes thoughtful conversation about pertinent issues. An offhand or misappropriated use of the term *resilience* does not overshadow its usefulness in shaping critical thought when it is applied in a more robust way.

I am an engineer researching post-disaster recovery, and I adopt social science methods in my approach. I am working on one case study in which a local government planned to build a more resilient wastewater infrastructure for earthquake-damaged communities. The government considered the repair of wastewater infrastructure to be a technical engineering problem and did not consult the community regarding its proposed solution. It perceived a need to expedite the construction of this infrastructure to facilitate community recovery back to a form of "normal" life and that external consultation could slow down the rebuilding process.

Some members of the community felt that the government's proposal was being imposed on them. Community opposition to the proposal ultimately led to a court case, and the government reviewed its decision. The government has now decided to repair the existing infrastructure in some areas rather than replacing it with different—more resilient—technology.

DANIEL ALDRICH'S REBUTTAL REMARKS

In my opening remarks, I argued that resilience—the capacity of community members to work collectively to cope with, adapt to, and transform after shocks such as disasters—is more than a buzzword. Instead, it underscores the power of social connections in helping people overcome adversity in industrialized and developing nations. Of course, governments, whether in the UK or elsewhere, would like to reduce their involvement in recovery; they often claim that communities should bear financial and administrative responsibility during a crisis. However, neither neoliberal arguments nor administrative desire to shirk responsibility reduces the importance of bottom-up, neighborhood, and community-based resilience.

Some observers who agree that resilience is important may just shrug their shoulders and argue that some communities are resilient and others are not. Fortunately, substantial research has demonstrated that communities can turn to social capital to improve their resilience to crises and disasters.[12] Several community-based programs have proven to have had a measurable impact in connecting residents and neighbors and increasing trust and reciprocity. One program is known as community currency or time banking.[13] Because so many people feel that time is limited, volunteering has been steadily dropping over the past decades. Community currency provides an incentive for residents to leave their homes and get involved in projects ranging from trash pickup to tutoring local schoolchildren. In return for volunteering, residents receive currency in the form of hours in a time bank, which can be exchanged at local businesses, at farmer's markets, or for bartered services. One hour of cleaning a river or mending a neighbor's fence might amount to five Toronto dollars. Studies have shown that time banking and community currency markedly increase trust and involvement and improve local business-to-business interactions in a virtuous cycle.

Communities hoping to increase resilience must take urban planning and the design of public space seriously. Too few towns and cities have sufficient space for recreation, social interaction, and leisure. Architects, such as those working with the Tohoku-based program iBasho, have come to understand that surroundings strongly shape social interactions; purposeful design of public housing, streets, parks, and piazzas can help residents interact and move away from the isolation that accompanies car-driven planning.[14]

Finally, several towns and cities, including Wellington, Tokyo, and San Francisco, have set aside funds to encourage the creation of neighbor-to-neighbor connections. These include the Neighbor Fest event where residents gather to meet and listen to music, *matsuri* (festivals) where Tokyoites dance and enjoy snacks, and sports days when local kids can release energy in a safe environment. Through face-to-face interactions, community residents can build ties that will help them navigate challenges and crises.

Local, regional, and national governments that recognize the power of people in mitigating and responding to a crisis should assist local communities in implementing these programs.

JONATHAN JOSEPH'S REBUTTAL REMARKS

It was helpful to distinguish between two approaches to resilience: one that considers it to have an intrinsic value, and the other that understands it as defined by conditions, context, interpretation, and application. This distinction is not so clear-cut and varies depending on the aspect of the term that is emphasized. However, it links to a comment discussing resilience as inseparable from social, economic, and cultural factors, one asserting that we must take into account the unequal distribution of these and other capacities and capabilities in our analysis of the concept.

Such a contextualized understanding of resilience points us in a more political direction and thereby problematizes the claim that resilience is all about the "human." The human is certainly a central element of resilience thinking; as I mentioned in my opening remarks, resilience recalibrates our understanding of the human and its capacities. Resilience does this by enabling a reflexive assessment of our embedded social context. It helps to develop a relational understanding of problems and raises questions about individual autonomy by emphasizing connectedness and social embeddedness.

Rather than representing a radically new idea, resilience revisits many of the issues raised in discussions of social capital. In drawing attention to specific elements of awareness, learning, adaptation, and recovery, discussions on resilience move beyond those about social capital. Resiliency scholars often invoke intangible human qualities: qualities associated with resilience and well-being are perceived as more human because they do not conform to rational-calculative capitalistic behavior. However, I argue that such qualities are consistent with neoliberal and market-based governance. They characterize humans as flexible, innovative, enterprising, and risk-taking. Such approaches, rather than replacing market logic, fill its gaps and ensure its continuity.

In fact, this is entirely consistent with a neoliberal perspective that quite rightly questions classical liberal assumptions about individual autonomy and rational planning by emphasizing the "messy" nature of social life, the embeddedness of economic activity in social norms and practices, and the inherently "wicked" nature of complex problems.[15]

I emphasize that resilience is not reducible to a neoliberal perspective. A more expansive approach looks at the variegated nature and definition of resilience. As noted in the comments, the UK approach to resilience is very centralized and prescriptive, but it invokes the idea of community in a different manner than countries such as France and Germany. I believe that the more Anglo-Saxon resilience approach tends to dominate; to say that the UK and institutions such as DFID and the World Bank are characterized by a more neoliberal approach to resilience implicitly recognizes that an Anglo-Saxon approach to resilience dominates policy areas like development and disaster risk reduction. Notably in this field, policymakers from the EU or Germany's Federal Ministry for Economic Cooperation and Development (BMZ) draw heavily on DFID's arguments on resilience. Somewhat vague notions like the turn to the "human" do not explain this process of political appropriation—hence the need to develop a strong political argument to render more meaningful alternatives to a neoliberal perspective.[16]

PARTICIPANT'S COMMENTS

Christopher Lyon, PhD student, Department of Geography and Environmental Science, University of Dundee, Nethergate, Dundee

Resilience perhaps becomes a neoliberal buzzword when it is misapplied in a setting that doesn't share the same ontological underpinning. For example, community social relationships are informal, complex, and dynamic, suggesting compatibility with the ontology of resilience. However, governance tends to be ordered in a linear and predictable manner and operate siloed institutions.

How resilience is understood is central to whether it is useful in the above-mentioned contexts. Perhaps resilience leans toward pluralistic interpretations, which I suppose signal complexity. But we are saddled with institutional cultures and politics that do not mesh with ambiguities inherent to complex organization and pluralistic ideas. Could it be that recognising context dependency and complexity in resilience means its definition and application is and must be inherently malleable? Could resilience then be rooted in rigorous questions about the who, what and how of each circumstance in which it is applied? The challenge inherent to debates on the status of resilience involves moving away from the reductionist thinking (bullet points, one size fits all definitions, metrics, and linear practices) that seems to be desired by policymakers and practitioners whether or not they are pursuing a neoliberal agenda.

As Jonathan notes, the complexity of lived reality presents a tremendous challenge to existing worldviews, in that it positions individuals, communities, and institutions as active constituents of experience, not just as entities that constitute its forms.[a] In practice, this means blurring lines between the governed and the government, and between disasters and their human and environmental causes, recognising that events do not happen in isolation.

[a] David Chandler, *Resilience: The Governance of Complexity,* (London and New York: Routledge, 2014).

DANIEL ALDRICH'S CLOSING REMARKS

Resilience, like other powerful concepts, can be made too vague to be useful and robbed of impact by governments and corporations looking to externalize costs. The idea of resilience is currently in vogue among academics and policymakers, but it would be a mistake for contrarians to prevent the idea from entering a tool kit of strategies to prepare for and adapt to hazards. Resilience recognizes the capacity for mitigation and transformation intrinsic to communities around the world. Decision-makers who take resilience seriously may emphasize how spending on social infrastructure costs less than traditional, large-scale investment projects, but this misses the point. Policymakers should invest in community resilience because it is highly effective, not because it reduces their financial or moral responsibilities. It would also be a mistake for policymakers to understand resilience as a bargain and think that it allows them to get something for nothing. Even if it is more cost-effective than building levees and seawalls, building social networks takes time, persistent presence in a community, and trust—none of which comes cheaply.

Successful resilience-building programs are bottom-up and community-driven. While many argue that governments must take on more financial and administrative burdens to handle hazards, top-down approaches cannot match the power of residents working cooperatively, as illustrated by James Scott.[17] Some of the best practices in the field of resilience have come from organizations like the Wellington Regional Emergency Management Office in New Zealand, which has become a go-to network as much for community development ideas as for disaster ones. Likewise, the Ofunato-based iBasho program in Japan has created a space with a library, cafe, and playground to reinforce and expand social ties among survivors of the March 11, 2011 triple disaster. San Francisco's Resilient Bayview program uses community planning to anticipate likely hazards and work communally to mitigate crises such as earthquakes and fires. The bayou-based community of Houma, Louisiana, has moved ahead of the federal government to enact a local sales tax to fund infrastructure protection from rising waters.

In these communities, residents, civil society groups, and faith-based organizations have not only physically and socially mapped hazards but also implemented programs to mitigate them. These neighborhoods are resilient precisely because they are not waiting for some government agency to step in to save them; the arrival of government agents postcrisis is invariably too little, too late, as Hurricane Katrina and the Fukushima nuclear disaster showed. Rather, these communities created a common vision of their future premised on neighbor-to-neighbor contact, trust, and collective action. Local communities understand their own needs, recognize which approaches will work, and commit to involvement for the long term. Further, communities can actively build trust and cohesion through approaches such as community currency, time banking, shared spaces, and open events.

Perhaps the most exciting application for resilience is that scholars now know that building resilience through deepened social capital and cohesion can serve as a platform for both climate change adaptation and disaster risk reduction. Many nations continue to pursue massive top-down investments in physical infrastructure, such as seawalls in Japan, levees on North America's Gulf Coast, and dykes in the Netherlands. But the most successful ways to handle wicked problems like extreme weather events, rising seas, and the destruction of ecosystems come from the collective action of communities and residents working together.

JONATHAN JOSEPH'S CLOSING REMARKS

In my opening remarks, I suggested two opposite positions on resilience: one that sees it as merely a buzzword; another that conceives it as a radical paradigm and an entirely different way of thinking and operating.

I would suggest that David Chandler's contributions tend toward the latter position.[18] For Chandler, resilience is so radically new that it disturbs all previously held ontological and epistemological assumptions. All that we know about ourselves and our place in the world is overturned by the assertion that we no longer live in a standard world of politics, science, or being human. I reject this view. Chandler's argument implies that these changes are the product of deep-rooted changes in society, nature, and humanity, rather than changing forms of governance.

Chandler also says that resilience is about self-governance and the constitution of a community. I agree with this stance. A political interpretation of this claim should emphasize that self-governance must be encouraged; it does not happen by itself. Communities are not inherently resilient; self-governance and community-building must be constructed and supported through strategic interventions, somewhat akin to what Michel Foucault understands as governmentality.[19] Some comments in this debate expressed a desire to move away from abstract discussions, but understanding the key factors invoked in resilience discourse—as opposed to accepting concepts like community, self-governance, adaptability, and human initiative as self-defined—is of vital political import.

I see resilience as a nonunique type of governmentality, consistent with recent trends for governing societies and their populations. I agree with those who say it represents a break from classical liberal approaches to governance, but I think this critique is consistent with neoliberal concerns in shifting responsibility from states to individuals and communities by championing active citizenship and private initiative. However, resilience implies adaptation, which functions as a form of self-governance in that it encourages people to change to confront risks and uncertainties, rather than try to change the world.

The value of resilience is evident, for example, in World Bank and USAID papers on disaster risk reduction and climate change in poor countries. They argue for adaptation rather than returning to traditional coping mechanisms. Such an approach involves changing the nature of societies and their economic activities. It places the burden of coping on the poorest communities while denying that their vulnerability is caused by structural inequalities.

PARTICIPANT'S COMMENTS

Nicola Musa, graduate student, Faculty of Social Sciences, University of Ottawa, Canada

Resilience has been defined in diverse and varied ways, but they all emphasize adaptability against stressors, disruptions, and adversity. Fran Norris looks at resilience as a process, not an outcome, that leads to adaptation.[a] She emphasizes the utility of stress and crisis to induce transient periods of dysfunction, which eventually lead to the adaptation of an altered environment. Ivan Townshend stresses the need for community members' voices to be heard and encourages public participation to build resilient communities.[b]

Over the past two decades, researchers have used a blend of quantitative and qualitative methodological research techniques to understand and measure cohesion and resilience, yet the concepts remain intangible and in need of further conceptualization and adaptation to apply to contexts of crisis or disaster.

The Syrian refugee crisis has had extraordinary social and economic effects on host communities in surrounding countries, including the reversal of development gains, stress on basic social services, and competitive labor advantages for the Syrians within limited employment markets. Jordan and Lebanon now have the highest ratio of Syrian refugees per capita worldwide (3RP: Regional Refugee and Resilience Plan, 2015). These traumatized communities have become permanently dysfunctional because their governments lack the necessary resources to deal with such stressors. As a result, the Syrian refugee crisis now poses a serious threat to regional and local peace, security, and safety in multiple communities.

[a] Fran H Norris et al., "Community Resilience as a Metaphor, Theory, Set of Capacities, and Strategy for Disaster readiness," *American Journal of Community Psychology* 41, no. 1 (2008): 127–150.

[b] Ivan Townshend et al., "Social Cohesion and Resilience Across Communities That Have Experienced a Disaster," *Natural Hazards* 76 (2014): 913–938.

THE MODERATOR'S CLOSING REMARKS: THE ETHICS OF RESILIENCE

Our debate confirms views that resilience is a malleable, abused, and fuzzy concept, yet remains useful in the face of disasters and climate change. The website of this online debate was visited 1,809 times by more than 550 people from about fifty countries. A total of 252 people voted on the question raised, and more than fifty-nine comments were recorded on the blog. We witnessed a major shift in participants' positions over the last few days of the debate. As many as 80 percent of participants supported the pertinence of the concept of resilience in pre-debate votes. By the end of the debate, 43 percent were no longer certain that the concept of resilience could be applied and mobilized without questioning its pertinence, value, and meanings. Yet, as we shall see, they rarely rejected the concept altogether.

The concept of resilience derives deep value from the contextual, dynamic conditions involved in its deployment and interpretation; in and of itself, the concept does not seem to hold an intrinsic value.[20] Even the originality and metaphorical significance of the concept were questioned.[21] Contrary to the original question proposed in the debate, results reveal that the concept can be useful even if it has been overused, and even if its meaning is malleable and slippery. In fact, the systemic attributes of the concept of resilience and its power to evoke the strengths and capacities of individuals and social groups created a moderate-to-high level of consensus among participants. Its capacity to energize hundreds of practitioners and academics worldwide to vote and enthusiastically write on the blog for more than eight days attests to its compelling status; arguably, very few concepts in disaster risk reduction would attract comparable interest.

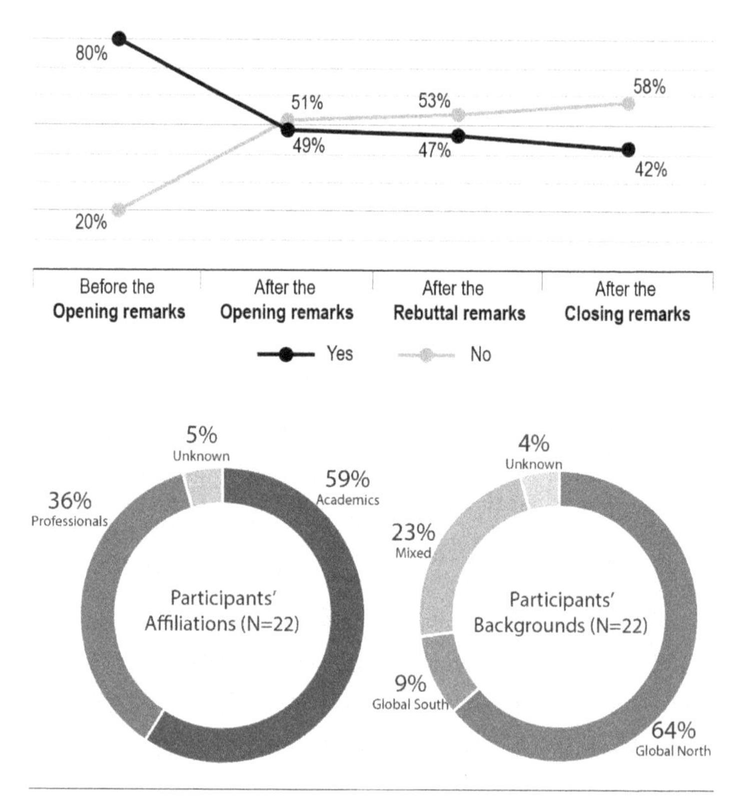

FIGURE 3.1 (*Top*) Results of votes at each stage of the debate. (*Bottom*) Distribution of affiliations and backgrounds among participants in the discussion.

Theoretical and practical implications emerged from the mixed results of the debate. From a theoretical point of view, voters and bloggers seem to urge academics to keep challenging, explaining, and questioning the role of resilience in academia, policymaking, and decision-making. From a practical point of view, debates on the concept of resilience raise two key ethical challenges: the first, to fully understand the ethical implications of adopting resilience as a concept in disaster risk reduction; and the second, to maintain an engaged search for stronger ways of expressing the need for, and desire to, create harmonious relations among nature, human beings, and the built environment. These are major tasks, but the interest kindled by the debate proves that a strong community is willing to face them. In times of climate change and crises caused by natural and manmade hazards, this is good news for societies and the environment.

NOTES

1. C. S. Holling, "Resilience and Stability of Ecological Systems," *Annual Review of Ecology and Systematics* 4 (1973): 1–23.
2. GFDRR, *Towards a Resilient Future: Annual Report 2012*, January 15, 2013, https://www.gfdrr.org/en/publication/annual-report-2012; Lee Bosher et al., "Built-in Resilience to Disasters: A Pre-Emptive Approach," *Engineering, Construction and Architectural Management* 14, no. 5 (2007): 434–446; Mark Pelling, *Adaptation to Climate Change: From Resilience to Transformation* (London: Routledge, 2011).
3. Sara Meerow, Joshua P Newell, and Melissa Stults, "Defining Urban Resilience: A Review," *Landscape and Urban Planning* 147 (2016): 38–49.
4. Sebastian Strunz, "Is Conceptual Vagueness an Asset? Arguments from Philosophy of Science Applied to the Concept of Resilience," *Ecological Economics* 76 (2012): 112–118; Fridolin Simon Brand and Kurt Jax, "Focusing the Meaning(s) of Resilience: Resilience as a Descriptive Concept and a Boundary Object," *Ecology and Society* 12, no. 1 (2007): 23–38; Karen I Sudmeier-Rieux, "Resilience: An Emerging Paradigm of Danger or of Hope?," *Disaster Prevention and Management* 23, no. 1 (2014): 67–80.
5. James Lewis, "Some Realities of Resilience: A Case-Study of Wittenberge," *Disaster Prevention and Management* 22, no. 1 (2013): 48–62, David E. Alexander, "Resilience and Disaster Risk Reduction: An Etymological Journey," *Natural Hazards & Earth System Sciences* 13, no. 11 (2013): 2707–2716.

6. Alexander Fekete, Gabriele Hufschmidt, and Sylvia Kruse, "Benefits and Challenges of Resilience and Vulnerability for Disaster Risk Management," *International Journal of Disaster Risk Science* 5, no. 1 (2014): 3–20.
7. Arif Mohaimin Sadri et al., "The Role of Social Capital, Personal Networks, and Emergency Responders in Post-Disaster Recovery and Resilience: A Study of Rural Communities in Indiana," *Natural Hazards* 90 (2018): 1377–1406.
8. Daniel P. Aldrich, "The Crucial Role of Civil Society in Disaster Recovery and Japan's Preparedness for Emergencies," *Japan Aktuell* 3, no. 2008 (2008): 81–96.
9. Daniel P. Aldrich, "Fixing Recovery: Social Capital in Post-Crisis Resilience," Purdue e-Pubs, 2010, https://docs.lib.purdue.edu/cgi/viewcontent.cgi?params=/context/pspubs/article/1002/.
10. Jon Coaffee et al., "Urban Resilience Implementation: A Policy Challenge and Research Agenda for the 21st Century," *Journal of Contingencies and Crisis Management* 26, no. 3 (2018): 403–410.
11. Gonzalo Lizarralde et al., "We Said, They Said: The Politics of Conceptual Frameworks in Disasters and Climate Change in Colombia and Latin America," *International Journal of Disaster Prevention and Management* 29, no. 6 (2020).
12. Daniel P. Aldrich and Michelle A. Meyer, "Social Capital and Community Resilience," *American Behavioral Scientist* 59, no. 2 (2015): 254–269; Daniel P. Aldrich, *Building Resilience: Social Capital in Post-Disaster Recovery* (Chicago: University of Chicago Press, 2012).
13. Bernard Lietaer, "Complementary Currencies in Japan Today: History, Originality and Relevance," *International Journal of Community Currency Research* 8, no. 1 (2004): 1–23.
14. Daniel P. Aldrich, "Trust Deficit: Japanese Communities and the Challenge of Rebuilding Tohoku," paper presented at the Japan forum, 2017.
15. Horst W. J. Rittel and Melvin M. Webber, "Dilemmas in a General Theory of Planning," *Policy Sciences* 4, no. 2 (1973): 155–169.
16. Jonathan Joseph, "Resilience as Embedded Neoliberalism: A Governmentality Approach," *Resilience* 1, no. 1 (2013): 38–52.
17. James C. Scott, *The Art of Not Being Governed: An Anarchist History of Upland Southeast Asia* (New Haven, CT: Yale University Press, 2010).

18. David Chandler, *Resilience: The Governance of Complexity* (London: Routledge, 2014).
19. Graham Burchell, Colin Gordon, and Peter Miller, eds. *The Foucault Effect: Studies in Governmentality* (Chicago: University of Chicago Press, 1991).
20. Juergen Weichselgartner and Ilan Kelman, "Geographies of Resilience: Challenges and Opportunities of a Descriptive Concept," *Progress in Human Geography* 39, no. 3 (2015): 249–267.
21. Fran H. Norris et al., "Community Resilience as a Metaphor, Theory, Set of Capacities, and Strategy for Disaster Readiness," *American Journal of Community Psychology* 41, no. 1 (2008): 127–150; Alexander Fekete, Gabriele Hufschmidt, and Sylvia Kruse, "Benefits and Challenges of Resilience and Vulnerability for Disaster Risk Management," *International Journal of Disaster Risk Science* 5, no. 1 (2014): 3–20.

PART II

TOUGH CHOICES

Tough Choices in Responding to Disasters: Flooding in Mozambique and California

LISA BORNSTEIN, TAPAN DHAR,
AND GONZALO LIZARRALDE

A TURBULENT WORLD

In March 2023, coastal zones on opposite sides of the world—in southern Africa and the western United States—were pummeled by extreme weather events.

In southern Africa, Tropical Cyclone Freddy hit Mozambique for a second time. Freddy had moved slowly, taking thirty-four days in a circuitous journey from Australia to Africa. Passing through the Indian Ocean, the tropical cyclone gained strength over the waters between Madagascar and Mozambique. It finally disgorged torrential rains, up to twenty-six inches in forty-eight hours, on parts of Mozambique, Zimbabwe, and Malawi. In both duration and overall energy released, Freddy broke meteorological records.[1] The cyclone's human impacts included upheavals to an estimated 1.07 million Mozambicans in eight provinces, with more than 184,000 people seeking refuge in emergency shelters. Flooding and landslides left more than 210,000 houses damaged and more than 50,000 people displaced.[2] While Freddy caused more than a thousand deaths in Malawi, 176 died in Mozambique where, importantly, trucks circulated with loudspeaker announcements advising people to go to higher ground; the advance warnings helped keep mortalities down.[3]

In addition to the direct impacts, international aid organizations point to how these recent floods, and associated displacement and damage to farmlands, worsened already dire problems in the affected regions of Mozambique. The country reflects a history of longstanding trade relations with India and the Arab World, Portuguese colonization, a ravaging sixteen-year civil war between an elected socialist government and armed opposition, and subsequent integration into the global capitalist system.[4] This history, and ongoing uneven development, is expressed in:

- The extreme poverty, below the 2014 global threshold of $1.90 per day, in which 60 percent of the population live
- Ongoing cholera outbreaks, high rates of HIV infection, and AIDS as the country's leading cause of death
- Drought-related food scarcity and hunger for the 90 percent of the country's people who rely on subsistence agriculture or informal work
- The lingering damage done by the devastating 2019 cyclones Idai and Kenneth
- In northern Mozambique, lack of shelter, basic services, food, and clean water for the estimated 730,000 to 1,000,000 people displaced by violence in that area[5]

In the northern gas-rich province of Cabo Delgado, localized conflict is deemed to constitute a complex humanitarian emergency. As in Afghanistan, Haiti, Myanmar, South Sudan, Syria, Yemen, and some other dozen countries worldwide, violence and insecurity frustrate local, national, and international responses to human need. In such complex emergencies, calamities—such as the drought, cyclones, floods, and crop failures that Mozambique has experienced in recent years—combine with conflict and failures of governance to threaten human well-being, safety, and survival. Yet even in stable parts of the country, extreme hardship is the norm for many people. According to one rural resident of central Mozambique, the continued hardships and lack of necessities mean that peace has not been achieved, even within areas free of armed conflict; in his words, "this is a new war." In this sense, Freddy is just one among many cascading challenges that require in-depth understanding, complex planning, and difficult decisions as to where and how different actors should respond.

UNDERSTANDING TRADE-OFFS IN RESPONSE AND RECOVERY

The complexity of conditions, even within a single country, means decisions do not emerge from an easy or simple set of technical calculations around emergency response and recovery. Each possible choice of action involves trade-offs. Whose voice and perspective will count? Which problems will be prioritized? Who will be responsible for which actions? How will immediate needs be balanced, or not, with longer-term ones, and with what implications for resource allocations and shortfalls? Who will benefit most and least at each point in time? And what can be done where aid is not easily provided, as in areas experiencing complex emergencies?

The case of temporary shelters is illustrative. Mozambique's National Institute of Disaster Management had activated its early warning systems, now well-established in the country, and made accommodation available in churches and schools. In the weeks after Freddy hit, some fifty thousand people were housed in these facilities. Food vouchers were distributed to those in the accommodations and to those who left them after the flooding subsided. But how long should such shelters be maintained?

The stance of the government of Mozambique was that temporary shelters heightened the risk of cholera outbreaks. There were nearly 11,500 cases of cholera in the weeks after the floods, including a rise in cases in the shelters. The government pointed to these numbers to show the wisdom of closing temporary facilities quickly.[6] Of course, there were other compelling reasons to do so. Previous crises had been met with the establishment of encampments that sometimes persisted for years. With Freddy, the shelters were in places where other activities of local importance occur. Since students needed to return to their classes and parishioners to their prayers, temporary accommodation was likely to be, indeed, temporary. In addition, for the government, timely closure of shelters marks a transition from direct cash transfers and humanitarian assistance to recovery and rebuilding, important when available funds are limited. If done swiftly, getting people back to their places of residence, or encouraging them to find alternatives, can be done while international humanitarian organizations are present and able to assist.[7] In the case of Freddy, organizations such as the World Food Programme and Care International were on hand and could explore ways to coordinate, finance, and carry out the transport of households toward home.

For cyclone-affected households and some aid organizations, the government's closure of shelters seemed premature. What the Mozambican government saw as an advantage—the cessation of direct aid—was seen negatively by those with no immediate alternative sources of food or income; delivery of food rations and other essential services was made both more difficult and less likely. The World Food Programme's March 24th Situation Report identified another difficulty with closing shelters quickly: many affected people could not yet return home.[8] Many roads were not yet passable, and even if boat, air, or four-wheel drive transport were made available by rescue organizations, the "last mile" connections to get home could be extremely onerous. More fundamentally, thousands of people no longer had a residence to which they could return. In Mozambique, the government addressed the problem by providing some affected households with construction kits. Yet, as the Mozambican historian and social analyst Bruno Mendiante observed, getting a kit one day and being expected to occupy the constructed house the next was like "using a sieve to cover the sun," an ineffectual measure that does more harm than good.[9]

UNDERSTANDING LOCAL PRACTICES, POSSIBILITIES, AND PITFALLS

There may, of course, be alternatives to temporary shelter. In Mozambique, some coastal residents have family in the highlands. Others have community or religious networks on which they can rely. Still others, especially those in the northern conflict zone, may seek safety in Tanzania or Malawi, though these neighbors fear an expansion of violence across their borders, which translates into a reluctance or refusal to welcome refugees. Understanding the possibilities open, or closed, to different affected households takes communication and time. But it also may generate knowledge of the locale—of how people manage their lives, their networks, their attachments to place, and the ways they approach daily joys and difficulties—that is crucial in emergency response, recovery, and risk reduction.

In a rapid assessment of the 2019 Idai floods, for example, the researchers Santiago Ripoll and Theresa Jones documented tactics used by central Mozambican households to deal with cycles of floods:

> Granaries are kept on stilts. People and cattle will seek refuge on higher ground, leaving people behind to make sure that harvest and assets are not stolen. There is mutuality between lowland and upland communities, with family members offering shelter when floods occur. It is not uncommon to have two plots of farmland (*munda*) to adapt to seasonal flooding, one in lower land close to the river (*matoro*), and another one in the uplands (*machamba*). Housing is also made out of grass and wood, that is more resistant to floods and can be replaced more affordably and with more ease than cement.[10]

Incorporating participatory inputs and community feedback, as seen here, can help emergency and risk reduction efforts to work through, with, and beyond local power structures.[11] For instance, households with reciprocal agreements around upland/lowland shelter might benefit more from transport vouchers than temporary shelter. Greater understanding of how aid can be diverted, who manages the allocation of food at the household level, or how local belief systems explain the causes of disasters can help international actors assess different options for aid.[12] Ideally, such insights can help to identify pathways toward recovery, development, and risk reduction strategies that promote human welfare, protect the rights of those most vulnerable (including women and girls), and build on, rather than replace, local initiatives.

Of course, participation itself may be an object of dispute and contestation. As shown in prior work on central Mozambique, rural people may associate participation with past requirements for work on community projects, and officials may see it as spaces for opposition groups to challenge administrative procedures.[13] A crisis may offer possibilities for change, for the type of transformative "building back better" often cited in disaster literature, but crises are also when established power brokers can consolidate their influence. How can aid be provided in ways that prevent capture by a few individuals? How can efforts to understand local realities move beyond reliance on selected local leaders who, when treated as interlocutors of local concerns, may reinforce prevailing inequitable social dynamics?

More broadly, the case of Mozambique points us to an important question, one relevant to the second case discussed below: What do complex disasters allow us to understand about historic patterns of disadvantage and the possibilities for their reversal?

WE'RE IN FOR NASTY WEATHER

That same month of March 2023, United States meteorologists were focused on the imminent arrival of the sixth severe atmospheric river of 2023 in drought-ridden California. Picking up vapor over the tropical waters of the Pacific, these atmospheric rivers stretch for sometimes thousands of miles, carrying important moisture to other parts of the globe. They can also bring massive precipitation and associated flooding, mudslides, and power outages to places like California, a state that, according to recent research, had just experienced its driest twenty-year period in 1,200 years.[14] Compacted soils and slopes denuded of vegetation by drought and wildfire were ill-equipped to absorb heavy rains. The strong winds, torrential rain, and heavy snowfall of the atmospheric rivers resulted in devastation, including fallen trees, collapsed highways, landslides, breached levees, and flooding. Emergencies were declared in forty-three of the state's fifty-eight counties. As one example, the U.S. Geological Service recorded more than six hundred landslides in California in the first three months of 2023.

The atmospheric rains transformed parts of the state's Central Valley. The drained bed of Tulare Lake—in the ancestral lands of the Native Yokut people—found itself hosting a 120-square-mile lake. Two hundred years of engineering, dams, levees, and canals had left the 750-square-mile lakebed suitable for the many towns, farms, prisons, sewage treatment plants, and other facilities now located there; all of these were at risk. Though the resuscitated lake attracted birdlife and, to the dismay of local officials, some boaters, thousands of acres were under water, with alfalfa, cotton, almonds, and pistachios among the crops lost. Fears were that the thaw of an atypically heavy snowpack in the state's Sierra Nevada mountains—locally labeled the Big Melt—would expand the lake, perhaps to larger than two hundred square miles.[15] Impacts would be felt at local, regional, and wider scales, with evacuations of people and livestock, further destruction of buildings and infrastructure, and a chance that sewage sludge would contaminate agricultural lands. As King's County sheriff David Robinson explained, "We're going to have a million acre-foot of water covering up an area that feeds the world."[16] Efforts to raise levees, put sandbag barriers in place, and divert river flows were underway.[17]

WHEN THE LEVEE BREAKS: RACE, ETHNICITY, INDIGENEITY, AND CLASS

Closer to the coast, Pajaro, an agricultural community in inland Monterey County, was also heavily hit. A nearby levee failed, releasing water through the town of three thousand residents.

Homes and businesses were knee-deep in floodwater. The National Guard came to help evacuate residents and to ensure that people stayed away from the crisis zone. It is an all too familiar story of an environmental disaster. But it is worth looking deeper to investigate what is sometimes termed the "dark side of planning."[18]

In the case of Pajaro, years of institutional neglect and marginalization of the town's low-income residents contributed to the area's vulnerability.[19] County boundaries set in 1850 split the Pajaro Valley in two, placing neighboring Watsonville and Pajaro in separate counties. Pajaro became home to those who were excluded elsewhere in the state, first to Chinese, then Filipinos,[20] and then, beginning in the 1940s, Mexican laborers. Historian Sandy Lyndon notes that "Monterey County didn't do the streets, didn't do plumbing, they didn't do anything in Pajaro. There was no infrastructure."[21] As of 2022, most of the unincorporated town's population were farmworkers, many of whom spoke only Spanish or an indigenous language such as Mixteco, Triqui, or Zapoteco.[22] Informed observers say these features of the population led the regional government to ignore the area. The president of the county board, Luis Alejo, observes that "in a county where 70 percent of the population [are] people of color . . . there is a deep history of marginalization."[23]

Such patterns of neglect are linked to the town's exposure to the 2023 flood hazard and the population's travails. The town had experienced four prior levee breaches—in 1955, 1958, 1997, and 1998—that prompted design of a still-to-be-realized flood management project.[24] Stu Townsley, a deputy district engineer at the Army Corps of Engineers, said that "in terms of 'benefit-cost ratios,' it never penciled out"; the low value of the properties in the area did not justify the $400 million price tag associated with the flood management upgrade.[25]

In the aftermath of the flood, residents of Pajaro settled into local hotels, stayed with acquaintances fortunate enough to have shelter, or went to county-provided shelter at the local fairgrounds. Nonprofit organizations and county agencies helped, supplying food, clothing, interpreter services, and assistance in getting financial support.[26] Expectations were of a long wait to return home. Most of the nine hundred inundated homes were deemed structurally sound, but water damage meant they were not safe for occupation. The houses and their contents—"refrigerators and stoves, family photos and electronics, mattresses, piles of clothes and children's toys"—were "contaminated with river muck" and in need of removal.[27]

Residents again found themselves at risk of falling below the economic thresholds to qualify for government support. A major disaster declaration from the U.S. president is needed to trigger direct aid—food aid, housing assistance, and health and legal services—to affected people. Brian Ferguson, the director of crisis communications for the Federal Emergency Management Agency (FEMA), explained that the criteria for disaster declarations include showing significant total monetary losses. In areas with high property values, damages "quickly add up to the totals they need" to qualify; in Pajaro's case, low home values, localized effects, and nonstructural damages made reaching the threshold difficult.[28] The state governor's eventual request for a disaster declaration grouped Pajaro together with farmlands and other affected municipalities. Heavy media coverage of Pajaro may have helped.[29] Based on combined damages from the atmospheric storms in eight counties, President Biden approved a declaration, allowing affected residents to apply for FEMA assistance.

Such patterns are replicated in other places and at wider spatial scales. Disasters—floods, fires, landslides, and earthquakes among them—can and do occur in wealthy parts of California. Loss of life, property, and livelihoods and associated trauma and hardship for survivors are always of concern. However, when disasters occur in areas with multimillion-dollar homes, FEMA funds are more easily mobilized. Wealthier owners are also more likely to have insurance; renters are the least likely to have insurance to cover disaster-related losses.

TAILORING, TINKERING, TRANSFORMING

As with the human impacts of Mozambique's cyclone, events do not occur in isolation. Vulnerabilities are built over time. While Californians are much wealthier than Mozambicans, dynamics of privilege, exclusion, and economic bottom lines are apparent in both situations.[30] Efforts to provide aid quickly and appropriately, and to do so in accountable and systematic ways, are also visible. Procedures allow for clarity and transparency but may, as these cases suggest, lead to regulatory barriers that reinforce vulnerabilities.

The 2023 storms described here had ecological, human, built, and economic impacts. Types of disasters, and the conditions in which they occur, can vary greatly. Different conditions demand different responses, but all cases require making difficult choices. For instance, with the complex humanitarian disasters mentioned above, conditions are often outright dangerous, both for those living in them and for any possible outside actors. For some observers, a decision not to intervene in a conflict zone reflects common sense and care for staff and volunteers who would be at risk; for others, it is a moral outrage that such tragic conditions persist, with calls for international action and demands that state authorities "unable or unwilling to cope" step up.[31] Likewise, for those displaced from such situations, high levels of uncertainty accompany the provision of aid. When will refugees and other displaced people be able to return, if at all, to their place of origin? Is the accommodation for those displaced to be for a few weeks, a few months, a few years, or far longer? Should priority be placed on establishing a safety net, fostering self-reliance, or a combination of the two? Should integration into the host region's society and economy be pursued or avoided?

In such cases, scarcity of facilities, logistic support, and relief materials, together with limited windows for international aid and fundraising, are factors in the often intricate and difficult decisions required in the aftermath of a disaster. Given human need and the limits of available resources, should public and international funds be heavily invested in that immediate emergency response phase? Or should resources and efforts be directed as quickly as possible towards the recovery phase?

Timelines for intervention also merit some thought. Disasters usually lend an urgency to disaster response so as to address people's pressing needs. Those needs are often met by humanitarian organizations, government agencies, and regular people stepping up to assist, whether with search and rescue, firefighting, medical treatment, food, clean water, warmth, temporary shelter, counseling, or monetary support. The immediate aftermath of a disaster is when media coverage is greatest and funds can be most easily raised. But other activities occur over longer timescales, and initial responses are not always well linked to such longer-term recovery and "building back better" efforts.

The debates in this part of the book tackle, among others, these questions of how to respond in the aftermath of calamities. In each chapter, debaters and commentators draw attention to understanding the contours of the disaster. The complexity of each situation, the commonalities with and distinctness from similar ones elsewhere, is important. So too is attention to associated timelines, resources, and power dynamics. Who is at risk? At what time frames? What are the capacities for government and humanitarian emergency response, aid, and recovery interventions, and under which organizational norms, safety protocols, and resource constraints? What alliances, rigidities, and power grabs are implicated in choices about appropriate responses? And what do all these mean for the voices, livelihoods, and life chances of displaced and other affected people?

The debates highlight tensions, trade-offs, and tough choices around such questions. Often contributors stress the need for responses, tailored to each situation, that can establish temporary safety nets. But they also stress that such approaches may simply tinker with a situation, doing little to transform underlying sources of risk or inequities. As you will read, some debaters call for changes within the prevailing system of disaster response; others see the system as so flawed—entrenched, for instance, in neocolonial relations that reinforce, rather than reduce, risks—that a fundamental and radical realignment is required.

THE DEBATES

The debates in this section explore how to act in the face of crisis, whether to do so with or without participation or aid, and how to fix what does not work well, fairly, or transformatively to reduce risk. They build on the recognition of collective responsibilities to respond to human needs with specific ideas about how to do so. They explore diverse dimensions of, and perspectives on, the construction of risk and vulnerabilities, the roles and limits of different actors in responding to calamities, and trade-offs and uncertainties in making choices.

Many experts believe that sheltering disaster victims and displaced people in temporary settings is unavoidable in the recovery process. However, studies show that temporary and substandard shelters suffer from several drawbacks and cause multiple secondary effects. What sociopolitical dynamics affect the early phases of disaster reaction? Does temporary housing hinder the recovery process? In chapter 4, Ilan Kelman and Graham Saunders examine standard answers to these questions and highlight the varied decision-making processes that determine how to treat disaster victims. Kelman argues that, by definition, temporary housing is insecure and thus at odds with efforts at stabilization and recovery. Saunders contends that temporary housing is an important complement to efforts at recovery and reconstruction. Debaters use examples from Haiti to develop their arguments. Both agree that, too often, temporary shelter persists and that there is inadequate support for longer-term recovery. The debate closes with presentations of institutional impediments, policy weaknesses, and possible policy advancements.

Chapter 5 offers a continued look at temporary shelter, with a focus on spatial dimensions: should people be sheltered in encampments or dispersed among the wider population? Kamel Abboud and Jeff Crisp examine the benefits and drawbacks of sheltering refugees in contained camps. If housing is to be provided, should refugees be sheltered in organized camps, or should they be free to find sheltering solutions in residential markets and slums? Camps facilitate the distribution of aid and resources and protect refugees from violence and economic exploitation; they also limit independence and access to jobs, economic opportunities, and education, and frequently hinder the integration of refugees into host communities. Abboud points to context and what is meant by encampment as part of his arguments in favor of temporary shelter. Crisp cites the persistence of camps throughout the decades, opportunities lost by investing in temporary facilities, and abridgements of the rights of those so housed as a preface to a call for change. Both the debaters and commentators draw out weaknesses of encampments, suggest policy reforms, and identify promising directions in international practices.

Practitioners often argue that participation is key to achieving disaster risk reduction and effective disaster response. But to what extent is participatory decision-making indispensable, and how effective is it in disaster risk reduction? Can existing power structures ensure equity in participation? In chapter 6, we explore public participation. While most readers will assume, like our online participants, that citizen engagement is desired and important, our two debaters—Christopher Bryant and Camillo Boano—find themselves on different sides of the issue. Bryant calls for continued and expanded citizen participation around planning for disaster risk reduction, seeing it as an important aspect of local governance. Boano critiques formal mechanisms of participation as failing to create spaces where people can, through the exercise of citizenship, challenge prevailing power dynamics that are enforced by project leaders, technocrats, and politicians. Both debaters acknowledge that inclusion of ordinary (or "citizen expert") knowledge can aid in risk reduction. Both are concerned, as well, with the ethical stance of different actors in participatory processes. However, the chapter also raises questions. Can participation be made more effective? Can it lead, in the context of crises and trauma, to transformative visions and plans for the future?

Chapter 7 returns to the theme of aid. Most people associate disasters with emergency aid and the work of charities during reconstruction efforts. But how useful is disaster aid? The debate between Anna Konotchick and Jason von Meding explores whether aid can effectively assist people and places to avoid disasters and/or rebuild afterward. Konotchick argues that yes, aid is essential in attending to those at risk but can be better deployed. Von Meding, in contrast, sees aid as influenced by political agendas, colonialism, imperialism, and domination, embedded in an unjust system that creates, and benefits from, risk; for him, attending to the politics of disasters implies dismantling the existing aid system. Their debate outlines different pathways to a paradigm shift in aid, with additional arguments and examples supplied by online debate participants. It also shows two faces of aid: on the one hand, aid often bypasses legitimate governments and authorities, sometimes flowing into corrupt hands; on the other, some charities and humanitarian groups perform crucial work to avoid injustices and eliminate oppressive systems.

NOTES

1. NASA Global Precipitation Measurement, "GPM Long-Lived Tropical Cyclone Freddy Brings Heavy Rain and Flooding to Madagascar and Mozambique," March 17, 2023, https://gpm.nasa.gov/applications/weather/news/tropical-cyclone-freddy-brings-heavy-rain-and-flooding-madagascar-and-mozambique; Weather Channel, "Cyclone Freddy Sets New World Record as Longest-Lived Tropical Cyclone," March 13, 2023, https://weather.com/storms/hurricane/news/2023-03-06-tropical-cyclone-freddy-mozambique-madagascar-record; World Meteorological Organization, "Tropical Cyclone Freddy May Set New Record," March 10, 2023, https://public.wmo.int/media/news/tropical-cyclone-freddy-may-set-new-record.
2. World Food Programme (WFP), "Mozambique External Situation Report #8," March 24, 2023, https://www.wfp.org/publications/Mozambique.

3. Frank Phiri, Manuel Mucari, and Carien du Plessis, "Cyclone Freddy Teaches Deadly Lessons on Storm Warnings, City Sprawl," Reuters, March 20, 2023, https://www.reuters.com/world/africa/cyclone-freddy-teaches-deadly-lessons-storm-warnings-city-sprawl-2023-03-20/. Tigere Chagutah, Amnesty International's east and southern Africa director, notes that "Mozambique and Malawi are among the countries least responsible for climate change, yet they are facing the full force of storms that are intensifying due to global warming driven mostly by carbon emissions from the world's richest nations." Associated Press, "Cyclone Freddy Wrecks Malawi and Mozambique, Killing More Than 200 People," NPR, March 14, 2023, https://www.npr.org/2023/03/14/1163380089/cyclone-freddy-malawi-mozambique-flooding-mudslides.
4. Lisa Bornstein, "Planning and Peacebuilding in Post-War Mozambique," *Journal of Peacebuilding & Development* 4, no. 1 (2008): 38–51; Lisa Bornstein, "City Fragments and Displaced Plans in War Torn Mozambique," *Open House International* 32, no. 1 (2007): 16–28.
5. WFP, "Mozambique External Situation Report #8," 1; ReliefWeb, "In Wake of Cyclone Freddy, Amid Disruption to Critical Services, Cholera Surges in Mozambique," March 20, 2023, https://reliefweb.int/report/mozambique/wake-cyclone-freddy-amid-disruption-critical-services-cholera-surges-mozambique.
6. Joaquim D. Lequechane et al., "Mozambique's Response to Cyclone Idai: How Collaboration and Surveillance with Water, Sanitation and Hygiene (WASH) Interventions Were Used to Control a Cholera Epidemic," *Infectious Diseases of Poverty* 9, no. 68 (2020).
7. The response phase is focused on immediate reaction to the emergency and includes, among other possible measures, activation of an emergency operations center, evacuation of at-risk populations, search and rescue, flood control or firefighting, opening of shelters, provision of emergency rescue and medical care, and supply of shelter, warmth, food, and clean water to affected populations.
8. WFP, "Mozambique External Situation Report #8," 1.
9. Marceline Mueia, "Zambézia: Governo pressiona vítimas do ciclone Freddy," DW, March 23, 2023, https://www.dw.com/pt-002/zamb%C3%A9zia-governo-pressiona-v%C3%ADtimas-do-ciclone-freddy/a-65096825.

10. Santiago Ripoll and Theresa Jones, "The Context of Sofala and Manica in Relation to Cyclone Idai Response in Mozambique," UNICEF, IDS & Anthrologica, 2019, 1–2.
11. Ripoll and Jones, "The Context of Sofala and Manica"; Lisa Bornstein, "Peace and Conflict Impact Assessment (PCIA) in Community Development: A Case Study from Mozambique," *Evaluation* 16, no. 2 (2010): 165–176.
12. Lisa Bornstein, "Peace and Conflict Impact Assessment".
13. Lisa Bornstein, "Planning and Peacebuilding in Post-War Mozambique"; Lisa Bornstein, "Politics and District Development Planning in Mozambique."
14. A. Park Williams, Benjamin Cook, and Jason Smerdon, "Rapid Intensification of the Emerging Southwestern North American Megadrought in 2020–2021," *Nature Climate Change* 12 (2002): 232–234.
15. Shawn Hubler, " 'The Big Melt' Has Begun in California," *New York Times*, April 24, 2023; Terry Castleman, "See the Rebirth of California's 'Phantom' Tulare Lake in Striking Before-and-After Images," *Los Angeles Times*, April 6, 2023, https://www.latimes.com/california/story/2023-04-06/california-tulare-lake-storms-flooding-satellite-photos.
16. Quoted in Soumya Karlamangla and Shawn Hubler, "Tulare Lake Was Drained off the Map: Nature Would Like a Word," *New York Times*, April 2, 2023, https://www.nytimes.com/2023/04/02/us/tulare-lake-california-storms.html.
17. California Governor's Office of Emergency Services (Cal OES), Tulare Lake Updates – Sept 2023. Sept 23, 2023. https://news.caloes.ca.gov/tulare-lake-updates-september-2023/; Cal OES, "Inside Look: Tulare County Recovery Efforts Underway," April 20, 2023, video available at https://news.caloes.ca.gov/central-valley-recovery-efforts-the-return-of-tulare-lake/.
18. Oren Yiftachel, "Planning and Social Control: Exploring the Dark Side," *Journal of Planning Literature* 12 (1998): 395–406.
19. Susanne Rust and Ruben Vives, "How a Long History of Racism and Neglect Set the Stage for Pajaro Flooding," *Los Angeles Times*, March 20, 2023; Kirsten Rudestam, Abigail Brown, and Ruth Langridge, "Exploring 'Deep Roots': Politics of Place and Groundwater Management Practices in the Pajaro Valley, California," *Society & Natural Resources* 31, no. 3 (2018): 291–305.

20. Christina Ayson Plank and Meleia Simon-Reynolds, "Watsonville Is in the Heart: Documenting Histories of Transpacific Filipino Migration in the Pajaro Valley," *Pacific Arts: The Journal of the Pacific Arts Association* 22, no. 1 (2022).
21. Rust and Vives, "How a Long History of Racism and Neglect Set the Stage for Pajaro Flooding."
22. Ruben Vives and Melissa Gomez, "Amid a Disastrous Flood, Interpreters Are a Lifeline for Indigenous Farmworkers," *Los Angeles Times*, March 29, 2023, https://www.latimes.com/california/story/2023-03-29/amid-a-disastrous-flood-interpreters-are-a-lifeline-for-indigenous-farmworkers.
23. Rust and Vives, "How a Long History of Racism and Neglect Set the Stage for Pajaro Flooding."
24. Pajaro Regional Flood Management Agency, "Pajaro River Flood Risk Management Project," 2021, https://www.pajaroriverwatershed.org/projects/pajaro-river-flood-risk-management; Edwin Flores, "Decades of Levee Failures Amount to Disaster in a Mostly Latino Californian Town Hit by Floods," *NBC News*, March 28, 2023. HYPERLINK "https://www.nbcnews.com/news/latino/decades-levee-failures-disaster-latino-farming-town-rcna77009" \h https://www.nbcnews.com/news/latino/decades-levee-failures-disaster-latino-farming-town-rcna77009.
25. Stu Townley, as quoted in Suzanne Rust, "Before Disastrous Flood, Officials Knew Pajaro River Levee Could Fail but Took No Action," *Los Angeles Times*, March 12, 2023, https://www.latimes.com/california/story/2023-03-12/authorities-knew-the-levee-could-fail.
26. Vives and Gomez, "Interpreters Are a Lifeline for Indigenous Farmworkers."
27. Susanne Rust and Ruben Vives, "Why Is Pajaro Not Deemed FEMA Disaster After Massive Flooding from California Storms?," *Los Angeles Times*, March 23, 2023, https://www.latimes.com/california/story/2023-03-23/why-is-pajaro-not-considered-a-disaster.
28. Rust and Vives, "Why Is Pajaro Not Deemed FEMA Disaster?"
29. Guy Carpenter, "Post Event Report: California Atmospheric River," December 26, 2022 to January 17, 2023, https://www.guycarp.com/insights/2023/02/California-AR-02-08.html.
30. World Bank, "Gross National Income per Capita 2021, Atlas Method and PPP, World Development Indicators database, January 15, 2023,

https://databankfiles.worldbank.org/public/ddpext_download/GNIPC.pdf; see also https://datatopics.worldbank.org/world-development-indicators/.

31. Stefan Dercon, *Gambling on Development: Why Some Countries Win and Others Lose* (London: Hurst, 2022), 321.

4

ON EMERGENCY RESPONSE

Does Temporary Housing Hinder the Recovery Process?

WITH ILAN KELMAN AND GRAHAM SAUNDERS

Temporary housing programs recognize that immediate shelter is required through periods of post-disaster reconstruction, as the latter typically takes years to complete.[1] Temporary housing is thought to provide households with a safe, healthy environment that permits them to resume daily domestic activities and that offers privacy and dignity at a transitional moment.[2]

Supporters of temporary housing programs often point to two ideas. First, if temporary accommodations are sufficiently durable to last the several years it may take until long-term shelter is built, affected families will not have to move several times, enabling them to experience a certain sense of physical and emotional permanence.[3] Second, some building solutions allow households to upgrade or, alternatively, incorporate parts (e.g., materials, extra rooms, or storage areas) of the temporary structure into permanent ones.

Opponents of temporary housing offer several reasons for their stance. First, substandard temporary shelters often become permanent, perpetuating environmental, social, and economic problems.[4] Second, even when local materials are used, the production of temporary shelters is highly expensive considering their limited duration.[5] Third, land rights of individuals and communities are poorly addressed by temporary housing solutions, often leaving families settled but without legal status for indefinite periods.[6] Finally, temporary housing units are often built badly and located remotely, far from services and infrastructure.[7]

ILAN KELMAN: TEMPORARY HOUSING HINDERS THE RECOVERY PROCESS

Shelter fulfills four main needs: physical and psychological health, comprising protection from the elements as well as a feeling of home and community; privacy and dignity for families and communities; physical and psychological security; and livelihood support. For the sake of this debate, I suggest that temporary housing cannot fulfill the basic needs of shelter and thus hinders the recovery process.

In particular, the fact that housing is temporary immediately inhibits the development of feelings of security and a sense of community because families know they will eventually move and thus will not invest extensively into making temporary housing a home. Other needs of shelter can potentially be met in temporary housing, but they are rarely achieved. As with families, donors are unwilling to invest extensively in temporary housing because of its provisional status. Then when temporary housing is completed, donors are less likely to finance permanent housing because they see that disaster-affected populations are living in some form of accommodation, irrespective of whether or not it meets the four main needs of shelter.

Families in temporary housing find themselves in a form of purgatory, unable to fully settle in their residence and frequently lacking a clear or realized timeline to move into permanent housing. In the meantime, temporary housing uses land and resources, interferes with community activities (for example, taking up school playing fields), and requires infrastructure to operate, including water, sanitation, and public transport services. As donor interest in temporary housing wanes over time, families are caught between making their temporary housing livable, so that it fulfills basic shelter needs, and planning or pushing for permanent housing. Managing these two objectives tends to slow down the achievement of both, inhibiting disaster recovery.

In theory, there is no need for this sequence to play out; it is possible to provide temporary housing for the masses and simultaneously work on developing permanent housing to ensure an efficient transition. In practice, an efficient transition is uncommon. Instead, temporary housing too often becomes long-term and the recovery process is hindered.

The debate prompt only identifies a problem but does not proffer a solution. Alternatives that provoke fewer problems than temporary housing are needed. I advance a challenge for discussion: temporary housing hinders the recovery process, but what options are or could be better? Is temporary housing the worst choice for shelter in disaster recovery?

GRAHAM SAUNDERS: TEMPORARY HOUSING FACILITATES THE RECOVERY PROCESS

It has been well documented that, when possible, households affected by disasters will begin rebuilding and reconstruction processes immediately. Such efforts should be supported. However, temporary accommodations may be needed for an interim period because of factors such as the impact of disasters on the capabilities and resources of households; the safe removal of debris; local economic abilities and regulatory environments; safety and security considerations; and the need for replanning or relocation. I contend that in such cases any temporary short- or medium-term solutions should be considered as components of a longer-term recovery process aimed at providing contextually appropriate shelter. These components include support for hosting families, cash for rent, and material and technical resources to construct improvised shelters on the sites of original dwellings, as well as the provision of a built or prefabricated structure.

Unfortunately, temporary housing has come to be perceived primarily as the provision of a housing "product." The term "transitional shelter" was originally used to refer to local building technologies that could be upgraded over time to transition into permanent dwellings. It has now become synonymous with the provision of poorly constructed timber sheds and other impermanent construction. The assumption, often shared by providers of temporary solutions, that others would provide more permanent housing options and address structural vulnerabilities has favored the adoption of temporary approaches over the implementation of recovery processes. All too often, short-term housing absorbs investments that could contribute to ensuring more durable shelter.

What impact does this situation have on families who can safely return to the site of their original home but who need some form of temporary accommodation? What impact does it have on households that cannot immediately return to the site of an original home because of topographic changes, political instabilities, or other factors? How do urban dwellers who may have resided on the upper floors of a multistory building or rented housing adapt to urgent circumstances when immediate reconstruction and reoccupation is a nonexistent or uncertain option?

When immediate shelter needs can be met through reconstruction processes oriented to the long term, temporary housing can indeed hinder recovery. Decision-makers and practitioners bear responsibility for bringing an informed understanding of operational and regulatory contexts, initial responses, the needs and capacities of affected households, and how temporary housing requirements fit within longer-term recovery processes. A more nuanced approach is required to meet immediate post-disaster shelter needs while providing more durable housing solutions. It is not an option to leave immediate needs for shelter unmet in favor of investing in longer-term permanent housing that may not be available for occupation for weeks if not months. But the provision of temporary shelter assistance should complement rather than impede permanent construction processes, thus helping to mitigate the negative social and economic impacts of impermanent settlements.

ILAN KELMAN'S REBUTTAL REMARKS

Why is the potential to use temporary housing to support people post-disaster and into recovery rarely fulfilled?

The 2010 Haiti earthquake, unfortunately, was an ideal situation for temporary housing to support recovery efforts. An immense calamity brought the world's gaze to the area and attracted significant donor support and experts with knowledge and guidelines for disaster recovery.[8] Yet Haiti still experienced similar problems to those experienced in other disaster-ridden areas over prior decades, problems that had galvanized the development and publication of guidelines for recovery. Why?

Did Haiti suffer from a lack of donor support and financing? Money was not lacking in many recovery sites after the 2004 Indian Ocean and 2011 Japan tsunamis, yet similar problems manifested. Money is needed and appreciated for disaster recovery efforts, but issues stem from more than financial shortfalls. Did these areas suffer from a lack of local permanent housing policies? Or were problems in advancing recovery solutions inherent to temporary housing initiatives? Guidelines and training provided by recovery experts are clear that permanent housing policies should not be neglected over temporary housing fixes; nevertheless, they are often ignored. Perhaps a short-term focus on temporary solutions fails to create a bridge to more permanent policies.

Have further explanations not contained in the guidelines been ignored in research, training, and practice? While we can appreciate existing case studies and be grateful for successful projects, we continue to face challenges by poorly implementing temporary housing for political and institutional reasons that are difficult to overcome. Or does the nature of temporary housing intrinsically hinder recovery?

GRAHAM SAUNDERS'S REBUTTAL REMARKS

From the perspective of individuals or a household, "temporary" is not a preferred state regarding employment, residency, access to services, or housing. However, in the absence of a permanent solution to these needs, temporary or interim solutions may be acceptable or essential. This debate has highlighted that greater discretion is required in using temporary housing as a descriptive term or programmatic approach to meet post-disaster shelter needs. Temporary housing should be supported when no security or regulatory impediments prevent disaster-affected households from repairing or rebuilding their homes; it could include the erection of a temporary structure to be progressively upgraded over time until it becomes a permanent home. For households displaced across locations or borders and unable to return home or secure a location for permanent resettlement, temporary, interim housing solutions are the only available option.

Similarly, the term "recovery process" should be contextualized within standard processes of housing development, conceived before disasters and enduring after a formal reconstruction phase. For individuals and households, housing is an iterative development, changing depending on available assets, shifting household demographics, regulatory environments, cultural or social factors, and economic opportunities. Any post-disaster housing intervention should reflect the changeability of housing possibilities, among which interim or temporary solutions may be desirable. Any shelter intervention, whether the provision of a tent or the construction of a permanent house, requires adequate analysis of potential building risks and their management. Temporary housing is typically unable to address all shelter and settlement risks and vulnerabilities, exposing households to new or increased risks. Thus, short-term shelter interventions, even when conceived as part of longer-term permanent solutions, should address the management of such risks by affected households. Transitioning temporary interventions to permanent shelters remains the most effective approach to managing and reducing such risks.

ILAN KELMAN'S CLOSING REMARKS

Further research and work are still needed to address how to overcome the ills of temporary housing. Such work should focus on, first, how temporary housing often becomes permanent despite its inadequate conditions and, second, how an emphasis on temporary housing can devalue the need for permanent housing policies and practices. Post-disaster situations are not easy for those affected. These individuals and communities deserve full support for recovery, which requires abandoning long-standing approaches and models on a case-by-case basis, especially when they are inappropriate for the particular context.

GRAHAM SAUNDERS'S CLOSING REMARKS

This debate has highlighted both the wide-ranging understanding and implementation of temporary housing and the challenges in advancing recovery and reconstruction activities in the immediate aftermath of disasters. Temporary or nonpermanent shelter is often the only option for displaced households who cannot settle permanently in their new locations. Similarly, for households formerly located in places that require extensive cleanup or redevelopment, or in sites facing regulatory barriers to reconstruction activities, some form of interim accommodation is necessary.

I suggest that rather than focusing on the application—or misapplication—of temporary housing approaches, we should reevaluate the widespread hesitancy to support recovery and reconstruction activities from the outset. Media and public pressure were partially responsible for driving the recovery response to the 2010 Haiti earthquake and the extensive erection of predominantly timber sheds, which directed financial resources to provide homes rather than emergency shelters quickly.

Ongoing research and analysis by the International Federation of Red Cross and Red Crescent Societies (IFRC) and other humanitarian groups expose regulatory barriers limiting the implementation of post-disaster shelters; established building, planning, and approval processes, and the limited capacities of administrative authorities, are not directed to meeting the urgency and scale of recovery processes. In many countries, ministries responsible for post-disaster shelter assistance, which is often categorized as social welfare, remain distinct from those responsible for the technical oversight of housing and construction. Senior management and national representatives often do not adequately understand issues of shelter and reconstruction, while agencies and donors typically lack institutional recognitional. Such siloed processes and interests perpetuate risk-averse approaches that prioritize short-term emergency or temporary shelters over engagement with long-term, more complex reconstruction. Recovery decisions are further limited by a poor understanding of institutional legal liabilities, property rights and tenure, and financial investments and costs.

As evidenced by many national examples, enabling affected households to initiate the reconstruction or repair of their homes progressively decreases the need for government or agency involvement, as such households independently invest resources and resolve ongoing issues.

We should ask not whether temporary housing hinders recovery processes but why support for recovery and reconstruction is not a default response. To this end, we should also review the institutional impediments within governments, donors, and humanitarian agencies that have normalized the provision of temporary housing over other possible responses and solutions.

THE MODERATOR'S CLOSING REMARKS: THE ETHICS OF EMERGENCY RESPONSE

This debate captured the attention of a wide audience that included students, scholars, and practitioners. The debate website was visited 1,576 times by more than 631 people in more than thirty countries. As many as 121 people voted on the question. We received more than thirty-three comments, about 40 percent of them from experts and professionals with experience in temporary housing in crises. Most importantly, the debate elicited significant arguments about the role of public, private, and nonprofit organizations and affected families in the aftermath of disasters. Participants confronted ideas on the benefits and drawbacks of emergency responses, making connections to and sometimes highlighting the lack of consideration of the long-term well-being of affected populations.

Participants presented strong arguments regarding the positive and negative impacts of temporary housing. At least three times during the debate, the overall results shifted from a majority Yes to a majority No. By the end of the debate, the poll suggested that participants tended to agree that post-disaster

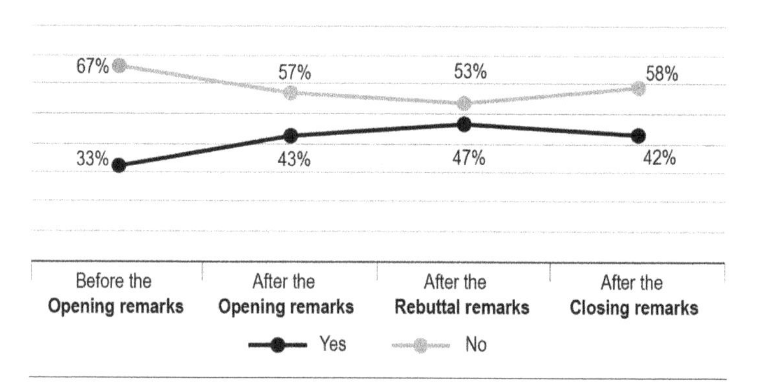

FIGURE 4.1 Results of votes at each stage of the debate.

temporary housing has more positive than negative effects. About 58 percent of participants responded No and 42 percent responded Yes to the question "Does temporary housing hinder the recovery process?" This does not mean that the drawbacks of temporary housing solutions should be ignored, but the results suggest that these negative effects should not overshadow the positive impact of temporary solutions in recovery processes. Graham Saunders captured the key advantages of temporary or transitional sheltering solutions.

This result makes Saunders's and Kelman's initial questions even more pertinent: How should we design, plan, and implement fast and efficient housing solutions that do not generate the negative secondary effects common to post-disaster recovery processes? How do we provide short-term solutions without reducing funding and detracting attention from permanent ones? How do we best link temporary measures to the sustainable ones required in medium- and long-term recovery processes? This debate has not provided a clear answer to these questions, but we believe it has provided significant analyses that may inform innovative solutions.

NOTES

1. Cassidy Johnson, Gonzalo Lizarralde, and Colin H. Davidson, "A Systems View of Temporary Housing Projects in Post-Disaster Reconstruction," *Construction Management and Economics* 24, no. 4 (2006): 367–378.
2. Cassidy Johnson, "Strategic Planning for Post-Disaster Temporary Housing," *Disasters* 31, no. 4 (2007): 435–458; Tom Corsellis and Antonella Vitale, eds., *Transitional Settlement: Displaced Populations* (Cambridge: Oxfam, 2005).
3. Daniel Félix et al., "The Role of Temporary Accommodation Buildings for Post-Disaster Housing Reconstruction," *Journal of Housing and the Built Environment* 30, no. 4 (2015): 683–699.

4. Daniel Félix, Jorge M. Branco, and Artur Feio, "Temporary Housing After Disasters: A State of the Art Survey," *Habitat International* 40 (2013): 136–141; Gonzalo Lizarralde and Colin Davidson, "Learning from the Poor," in *Post-Disaster Reconstruction: Meeting Stakeholder Interests*, ed. David Alexander et al. (Florence, Italy: Firenze University Press, 2006): 393–404; Krutika Gulahane and V. A. Gokhale, "Design Criteria for Temporary Shelters for Disaster Mitigation in India," in *Participatory Design and Appropriate Technology for Disaster Reconstruction: Conference Proceedings, 2010 International i-Rec Conference*, ed. Gonzalo Lizarralde et al. (Montreal: GRIF, Université de Montréal 2012).
5. Scott Leckie, *Regulatory Obstacles to Rapid and Equitable Emergency and Interim Shelter Solutions After Natural Disasters* (Geneva: IFRC, 2011); Cassidy Johnson, "Strategies for the Reuse of Temporary Housing," in *Urban Transformation*, ed. I. A. Ruby (Berlin: Ruby Press, 2007), 323–331.
6. Félix, Branco, and Feio, "Temporary Housing After Disasters."
7. "Alternatives to Camps," UNHCR, 2020, https://www.unhcr.org/alternatives-to-camps.html.
8. Corsellis and Vitale, *Transitional Settlement*.

5

ON SHELTER

Should Refugees Be Sheltered in Contained, Organized Camps or Be Allowed to Disperse Throughout Urban and Rural Areas?

WITH KAMEL ABBOUD AND JEFF CRISP

Whereas some specialists consider camps an effective solution that offers protection and essential services to refugees and forcibly displaced populations, others perceive them as places of control, segregation, and restricted movement.[1]

The former group typically argues that camps facilitate administrative tasks and aid distribution and help provide legal and institutional status for refugees. They also contend that settling refugees in camps allows host governments to address security concerns, attract donations, and distribute responsibilities among international humanitarian organizations such as the United Nations High Commissioner for Refugees (UNHCR). Defenders of camps also emphasize that those managed by international institutions help protect refugees from exploitation and marginalization.[2] Finally, they point out that by settling refugees in camps, governments and humanitarian organizations can reduce competition for limited services, affordable housing, and jobs, potentially minimizing tensions between refugees and local populations.[3]

Other scholars and practitioners critique camps as isolating refugees, limiting their access to labor markets, and hindering their social integration into new communities.[4] Despite initial planning and intentions, camps often become semipermanent or permanent shelters that are expensive for host governments and humanitarian organizations to maintain.[5] Critics typically blame camps for negative secondary effects on refugees' mental health and education and point to the camps' social and physical environments as facilitating crime, sexual abuse, and violence.[6] They believe that these drawbacks can be prevented by settling refugees in urban and rural areas; a dispersed approach can further refugees' socioeconomic integration within host communities. Moreover, they point to the importance of refugees' freedom to enter the labor market, find decent housing, and establish social ties with host communities—all of which are easier to achieve when settled in regular urban and rural areas.[7]

KAMEL ABBOUD: REFUGEES SHOULD BE SHELTERED AND CONTAINED IN ORGANIZED CAMPS

I would prefer to begin by reformulating the question of this debate to suggest a slightly different approach: "Under which conditions should refugees be hosted in urban and rural areas or sheltered in organized camps?" Indeed, I find it difficult to narrow all possible actions to two. How can we answer the initial question without putting it into context? Are all refugees alike? Are the conditions of all refugees alike? Are refugees running from a tornado in Florida considered to have the same needs as those of Haitian refugees after an earthquake that demolished an entire city? Or is the experience of migrants drowning by the thousands under the "vigilance" of some European coastguards considered the same as that of Ukrainian citizens fleeing Russian-backed militias in Crimea?[8]

Similarly, I am uncertain that "camp" has a universal definition. Does the term refer to a primary group of tents, shelters, or small housing units? If a camp is built with solid components, would it still be called a camp? Following that logic, when do camps become settlements, when do settlements become temporary villages, and when do temporary villages become cities? Is the definition of a camp based on its lifespan; that is, if a camp remains after fifty years of building adjustments and transformations, can it still be called a camp? Is the definition of a camp linked to the freedom of movement of refugees?

Before drafting any responses to my rephrased questions, we must consider several important factors.

1. *The cause of migration.* Is migration the result of naturally reversible phenomena (e.g., floods, fires, and hurricanes)? An irreversible disaster (e.g., global warming disasters and earthquakes) or disasters caused by human activities (e.g., nuclear disasters and dam failures)? An economic situation like extreme poverty or political turmoil and warfare that catalyze unpredictable outcomes in the short term?
2. *The needs and wishes of refugees,* taking into consideration their diverse cultural and educational backgrounds. Sometimes refugees reject specific hosting solutions. For instance, Palestinian authorities never accepted that their fellow refugees be hosted anywhere other than in provisional camps, so as to maintain political pressure in international relations. They reasoned that they were nurturing "the right to return" to their despoiled

land. Such policies have provoked multiple disasters in Jordan, Syria, and of course Lebanon, where refugees have been living in squalid camps for more than half a century.

3. *Spatial resources, territorial characteristics, and means of transportation.* Analysis of such elements is crucial to formulating an adequate hosting policy. How far are refugees allowed to move away from their homeland? Could they quickly return to their home country if the situation allows?

4. *The number of refugees relative to the inhabitants of the location*, whether a village, town, region, or country. For instance, consider how a hosting community of one thousand inhabitants could (or could not) provide aid, food, and shelter for a sudden flow of fifty thousand refugees, which is the case in several Lebanese villages.

5. *The characteristics and means of the hosting community.* Here we must include considerations of geography, climate, economy, politics, and, above all, the resilience of the existing infrastructure. This factor is especially important in the case of neighboring countries where political instability would jeopardize the security of refugees and local inhabitants.

We often do not have the time or luxury to conduct as thorough an analysis as is suggested above. Decent housing is the minimum acceptable condition to preserve human dignity. But is it possible in every situation? Are humans generous enough to provide that minimum for all refugees? As of today, the answer to this question is, unfortunately, no. NGOs must deal with refugee emergencies, often responding with the immediate solution of organizing camps where refugees can receive proper assistance and organized aid.

JEFF CRISP: REFUGEES SHOULD NOT BE SHELTERED AND CONTAINED IN ORGANIZED CAMPS

A common and largely uncontested practice for refugee aid throughout the twentieth century involved placing them in camps when they arrived in their country of asylum. The advantages of this arrangement appeared self-evident. For aid agencies, concentrating refugees in a single location made it logistically easier to provide them with the shelter, food, water, sanitation, and other forms of emergency relief that they required. Refugee-hosting countries thought that fewer security problems would emerge if refugees were kept in camps where their activities could be closely monitored. Accommodating refugees in large, highly visible settlements was generally thought to facilitate the raising of needed donor funds.

Over the past fifteen years, such assumptions have been increasingly scrutinized. Aid agencies such as UNHCR have sought alternative asylum solutions to camps, enabling refugees to live in urban centers or among host communities in rural areas. This major policy change, which I believe to be a very positive one, is the result of several factors.

First, while camps might be an appropriate temporary response at the height of an emergency, refugee situations usually last for years, or even decades. And throughout that time, camp-based refugees and their offspring have often been denied basic rights—such as freedom of movement, access to land and the labor market—and been unable to establish a livelihood.[9] Refugee camps are often located in remote, isolated, and inhospitable areas, making it impossible for refugees to grow their food and contribute to the local economy.

Second, even a well-resourced refugee camp provides an unnatural and often dangerous environment for its inhabitants, especially women and children. Camps are frequently characterized by high levels of sexual violence, the forced recruitment of adults and minors into militia groups, and attacks from hostile external forces. Encampment is also known to generate trauma, psychosocial problems, and intergenerational conflict, making it difficult for individuals and communities to prepare for a peaceful and productive future.

Third, even in countries where refugees are officially obliged to live in camps, growing numbers of refugees choose to vote with their feet. For example, refugees in Kenya have moved to urban centers like the capital city of Nairobi, which promises better opportunities to find work and live normal lives. At the same time, a growing proportion of the world's refugees are moving to countries where, for a variety of reasons, camps are not established or are used only to a limited extent: Egypt, India, Jordan, Lebanon, South Africa, and Turkey, for example.

Fourth, camps represent a lost opportunity. They make it more difficult for refugees to integrate with local communities and prevent them from acquiring the skills they will need to eventually return and contribute to the rebuilding of their own countries, should this be possible. Refugee camps absorb a huge amount of scarce donor funding without offering their inhabitants the opportunity to become self-reliant and find productive and permanent solutions to their plight.

Some situations might offer no other options than to establish refugee camps, especially in the initial stage of an emergency, but this is no excuse for camps to be supported for years, progressively losing the interest and resources of the international community. As UNHCR has recently recognized, alternative asylum solutions to camps must be pursued as a global principle, while the provision of appropriate support to refugees in urban centers and rural host communities must be improved.

PARTICIPANT'S COMMENTS

Faten Kikano, PhD candidate, Faculty of Environmental Design, Université de Montréal, Canada

Humanitarian agencies play a crucial and essential role in assisting refugees, especially in the emergency phase. Their support to host countries is also undeniable. However, they often act according to the policies and decisions of donor states; very few humanitarian agencies receive private funding (like Médecins Sans Frontières). Until 1997, refugees outside organized camps were not allowed to receive aid or protection, an injustice that stimulated internationally funded NGOs to advance the idea of camps as tools of donor states for "containment and control."[a] As the proverb states, "one man's meat is another man's poison," and a refugee population creates and sustains jobs in the humanitarian field. However, since we have only secondhand information on these issues, I suggest listening to Crisp's observations, as he is an expert who worked with the UNHCR for years. He could confirm, deny, or nuance such theories.

Studies show that refugee settlement policies are shaped by the political and economic interests of host countries. For instance, when Jordan hosted Iraqi refugees in urban areas, the refugees were not "countable," so Jordan was unable to receive its expected funding from the international community for acting as hosts. The Jordanian government learned from this experience and responded to the

Syrian crisis by creating camps, keeping the poorest refugees outside of the job market while promoting their aid enough to receive international funding. In Lebanon, the "policy of no policy" (open borders and no camps for Syrian refugees) pushed a large part of the Lebanese population below the poverty line, among other deplorable consequences. However, analyses demonstrate that this policy benefits the wealthy Lebanese who profit from a cheap labor market that exploits Syrian refugees.

[a] Guglielmo Verdirame and Barbara Harrell-Bond, *Rights in Exile: Janus-Faced Humanitarianism* (New York: Berghahn, 2005); Michel Agier, *On the Margins of the World: The Refugee Experience Today* (Cambridge: Polity, 2008).

KAMEL ABBOUD'S REBUTTAL REMARKS

The participants' comments are very relevant and help shift this debate. Among the possible directions are suggestions that the definition of "refugee" per the 1951 Geneva Convention is somehow obsolete and that "refugee" can refer to the case of displaced persons within the same country; the idea of providing camps with extra sites to support the development of self-sustaining communities; and the assessment that bringing assisted refugees into poor areas will only bring more poverty and despair. Such ideas merit more attention, but I will retain focus on the initial topic.

We agree that encampment is the first necessary step in helping refugees, but I contend that encampment should not be permanent. As Crisp highlights, the advantages of encampment appeared self-evident for aid agencies in the second half of the twentieth century. But after half a century of debatable results for the existing encampment model, especially in the Middle East, key aid agencies are reconsidering their initial approach to settlement solutions. The search for alternative settlement policies is urgent given the exponential growth of refugee populations and the inversely proportional dearth of funding. To be implemented, such new approaches must be developed in coordination with international politics, which, being politics of the mightiest, often does not seriously consider the opinions of the smallest and most vulnerable countries. Aid agencies will, of course, take into account humanitarian concerns and the conditions of the hosting countries, but they will align solutions with the interests of the powerful countries.

Concurrently, the world is witnessing a stiffening of asylum regulations in most Western countries, even those receiving a relatively small number of refugees. Western leaders and the UNHCR frequently declare intentions to integrate refugees within local populations, but they do not declare outright that such integration is intended to be achieved in countries adjacent to nations in conflict—to contain regionally the perceived refugee crisis.

Some weaker nations that lack sufficient resources to provide aid cannot manage the burden of the humanitarian crisis on their soil. Massive changes in population, cultures, and demographic equilibrium might have irreversible consequences, as in the case of Lebanon. According to the UNHCR, Lebanon hosts an estimated 1,835,840 Syrian refugees, while 295,000 Palestinian and 17,000 refugees live in camps, amounting to 2,147,840 refugees among a population of 4.4 million Lebanese people.[10] Integrating such a massive group of refugees—comprising 48.8 percent of the population—into local communities is not a viable solution.[11] UNHCR is helping as much as possible by providing $236 million of aid over six years out of the $2.75 billion requested, but this only amounts to less than twenty-two dollars per year per Syrian refugee. In such conditions, how could we predict any sustainable funding or planning solutions? Already more structurally, socially, and politically precarious than other nations, Lebanon is on the brink of economic collapse under the burden of its humanitarian crisis.

In my opinion, the option to "integrate refugees within the local population" might backfire in some situations and end up worse than a camp. Camps are a horrible way of sheltering refugees, but maybe their problem lies in their initial organization process. Until recently, most camps were planned and organized without urban and architectural considerations. It may be time for urbanists to be involved more deeply in the planning of camps, not merely by providing designs for housing but by offering more profound reflections about space and the nature of temporary dwellings.

JEFF CRISP'S REBUTTAL REMARKS

I agree that we should consider the original question of this debate in a more nuanced manner.

First, my opening comments were strictly limited to large-scale refugee situations in developing countries and were not intended to relate to the situation of refugees traveling by boat, to people displaced by environmental disasters in highly industrialized countries, or to asylum seekers who are in transit, moving from one country to another. The settlement options pursued must be tailored to the nature of the displacement that has taken place and the country's context.

Second, as Abboud has rightly suggested, the notion of a "camp" requires further interrogation, as this word is currently used to describe a wide variety of different situations: rural settlements where refugees have access to land and few limits placed on their freedom of movement; closed camps surrounded by fences to contain and confine the refugee population; informal settlements established by refugees, as in Calais; and

Palestinian refugee camps, which can be difficult to distinguish from the surrounding urban environment.

Third, as suggested in my opening commentary, it is important to consider temporal frameworks in assessments and implementation of encampments. An approach to refugee settlement that is appropriate and effective in the earliest stage of a mass influx will almost certainly have to be revised as the situation stabilizes. Such revisions are especially likely when the longer-term needs and aspirations of refugees, especially education, employment, and livelihoods, become of higher priority than the provision of emergency relief. At the risk of overgeneralizing, I maintain that refugee camps of the type traditionally established by UNHCR in developing countries make it more difficult for these needs and aspirations to be met, and thus make it more likely that refugee rights will be violated.

Finally, Abboud correctly reminds us of the need for refugee preferences to be taken fully into account when planning settlement solutions. Situations may arise in which refugees from the same country prefer to be moved to specific locations, either for security reasons or because camps provide the basic services required by the most vulnerable members of the community.

At the same time, it is important to recognize that camps have frequently been used by refugee elites and militia leaders to exert control over civilian populations and deprive them of any choice in their place of settlement. Such situations must be avoided if we are to prevent a repetition of the human rights violations perpetrated by the Khmer Rouge on Cambodian refugees in Thailand and Rwandan genocidaires in Tanzania, for example.

While I generally disfavor the maintenance of long-term refugee camps and consider it preferable for refugees to be allowed in urban areas and rural host communities, I acknowledge the dangers of these approaches. Once refugees are out of camps, a risk emerges that they will also be out of mind.

To avoid such scenarios, three things must be done. First, effective monitoring and outreach mechanisms must be established, allowing an ongoing assessment of the state of the refugee population, especially those who are least able to support themselves. Second, an active advocacy program must be designed to ensure that non-camp refugees can exercise the rights to which they are entitled, especially access to the labor market and freedom from harassment and exploitation. Third, if refugees are to be accommodated outside of camps, the needs of members of the host community must be met and their perspectives taken into account. Refugees cannot be protected adequately, nor can solutions to their plight be implemented, if they are unable to live in peace with their neighbors.

KAMEL ABBOUD'S CLOSING REMARKS

I would like to return to a major point raised by Jeff Crisp in his rebuttal. Crisp said his comments were "strictly limited to large-scale refugee situations in developing countries," which leads me to raise the perspective of a citizen of a "developing country" or a country that has submitted to the dictates of managing a refugee crisis and sheltering refugees on its soil.

When the situation becomes unbearable for host countries and refugees, only two alternatives typically are considered: (1) send the refugees back to their original country and try to relocate them to secured zones, supervised by UN agencies such as UNHCR; or (2) dispatch them, at any cost, to other countries, preferably industrialized ones. Since some countries have not signed the 1951 Geneva Convention protocol, with its binding "non-refoulement" article against sending people back to places where their lives are at risk, the application of such measures could be impossible.

I hope readers will agree with my tentative answers to the original question. First, encampment should be an immediate response, but such camps should not necessarily be contained. Instead, conditions should be revised to promote freedom of movement, self-supporting activities, ongoing assessment of the adequacy of housing, advocacy and control, and innovative organization of camps. A second step should involve monitoring rural or urban shelters once a final destination for refugees is identified, with attention to the refugees' culture, education, and native landscapes. Dispatching refugees in small groups to different regions or countries could help counter the ghettoizing effect of concentrated numbers of asylum seekers in a host society.

Changes to human civilization are rapidly taking shape. The future is unknown yet imaginably dark. Global warming, uncontrolled demographics, overexploitation of nature, and scarcity of resources will harshen living conditions. The aftermath of globalization will leave less affluent populations in an increasingly precarious situation. People will circulate across the world parallel to the circulation of goods, from the poorer Global South/East to the wealthier Global North/West. If no revolutionary vision for the future of humanity is conceived, we will experience decades of global impoverishment.

In my opinion, major human migration waves have not yet begun. To date, we have only witnessed preliminary population displacements. Asian populations, mainly Chinese and Indian people, have not yet been compelled to migrate outside their respective borders or, if moving, are doing so at a slow and peaceful pace. International organizations, as currently structured, are not prepared to handle the consequences of such mass migrations. We have no choice but to drastically change our vision for the future.

Poverty, violence, and wars will multiply exponentially at a global scale unless we find ways to create a new utopia, implement an economic system to supplant unbridled capitalism, control the international policies of countries exempt from the UN veto system, and dismantle the lobbies of the armaments industry and those of major financial groups.

International authorities and agencies should focus on developing new approaches to conflict prevention and not just post-disaster assistance. Preventive political, economic, or even, in the worst case, military action would be more effective and less expensive than current post-disaster relief aid.

JEFF CRISP'S CLOSING REMARKS

Kamel Abboud and I are in clear agreement that "camps are a horrible way of sheltering refugees." As I have suggested before in this debate, camps are usually associated with violations of fundamental human rights, especially with respect to freedom of movement. They are often located in hostile environments and have rarely provided their residents with decent living conditions or the opportunity to become self-reliant.

By definition, camps segregate refugees from other members of a community. They do so based on a misguided assumption that if refugees are allowed to mix and integrate with locals they will never choose to return to their countries of origin. In fact, for the many thousands of refugee children who have been born and raised in refugee camps, it is difficult to determine their so-called country of origin.

I fully agree that if it is necessary to establish a refugee camp in a given context, then it should be designed and managed as humanely as possible, drawing upon the insights of urban planners and architects. As Abboud has correctly stated, there have not been sufficient efforts to draw upon such expertise in the past.

Doing so will not be easy, given limited resources and the reality that many of the world's refugees are found in dysfunctional or authoritarian states. We cannot expect refugee camps to be prosperous and democratic when they are in such countries. At the same time, we must acknowledge that a camp, once established, is very difficult to dismantle. But that is what is needed to allow its residents to live a more normal way of life.

On a more optimistic note, a paradigm shift seems to be taking place in the international approach to refugee settlement and assistance. For many years, international communities normalized confinement of refugees to camps and, to provide for their basic needs, reliance on long-term "care and maintenance" programs spearheaded by UNHCR and other humanitarian agencies. This standard model of refugee aid is no longer sustainable. As UNHCR statistics demonstrate, most refugees no longer live in camps, either because the host state has chosen not to establish camps or because refugees have chosen to live elsewhere.

Rather than providing refugees with goods, it is becoming the norm to provide them with cash so that they can make their own purchasing decisions. The notion of "care and maintenance" has been discredited, and more attention focuses on how refugees can be helped to find decent work and establish their livelihoods. Largely due to the massive influx of Syrian refugees into host countries such as Jordan and Lebanon, the need to pursue area-based approaches that take full account of the needs of host communities is more widely recognized.

After decades of discussion, development actors such as the World Bank and UNDP are finally engaging with the issue of human displacement and are promising to address the challenge of providing refugees with appropriate and equitable long-term settlement options.

In the 1990s, I visited a camp for Liberian refugees in Sierra Leone. I came across a young man who had finished constructing a rudimentary mud hut to replace the tent he had been provided. He picked up a wooden stick and on the wall of the hut inscribed the ironic words, "This is not my home." A refugee camp can never be a real home, and we should do everything possible to avoid their establishment.

PARTICIPANT'S COMMENTS

Julien Deschênes, graduate student, Université de Montréal, Canada

If wealth were adequately distributed within and between nations, livelihood opportunities and housing solutions would not put undue pressure on vulnerable local populations. In a perfect world, I would advocate for formal work and housing opportunities provided to refugees, but reality has proven that migrants tend to be hosted in more informal and accessible solutions. Informality, as chaotic as it may seem, creates the networks and independence needed for successful integration processes.

To overcome this urgent issue, we need more than money. We need time, infrastructure, ideas, and most importantly, a shift toward a more humanistic general mindset. Our efforts concern helping one another in difficult times. Economic, political, military, and religious gains are far less important than the promise of the UNHCR to ensure that everybody has the right to seek asylum and find safe refuge when fleeing violence, persecution, war, or disaster at home.

THE MODERATOR'S CLOSING REMARKS

This debate has turned into a rich and lively conversation that attracted a sizable audience over a week's time. The debate webpage was visited 2,100 times by 826 people from eighty-two countries, with more than 350 people voting on the question under consideration. Moreover, the debate attracted thirty-five comments by students, scholars, and practitioners.

Beyond this quantitative assessment, the diverse and thought-provoking arguments put forward by panelists and commentators about the benefits and shortcomings of the encampment and non-encampment approaches remain the most important outcome of this debate. These comments played an important role in widening the terms and questions of the debate. Participants widely agreed that refugee settlement is an inextricably contextual problem, remaining multifaceted and becoming increasingly complex. It is not only unrealistic but counterproductive to look for simple, one-size-fits-all solutions.

As panelists and commentators exchanged views and ideas, a poll was conducted on the question "Should refugees be sheltered and contained in organized camps or urban and rural areas?" In the pre-debate period and during the early days of the debate, voters leaned toward the no-encampment option. But as the debate unfolded, the poll tightened and the pro-encampment option ended with a slight majority (58 percent).

I believe this shift in opinion reflects the increasingly nuanced views that panelists and commentators developed about encampment and non encampment approaches. Over time, participants and panelists tended to converge and agree on two key ideas neatly summarized by Abboud in his concluding remarks:

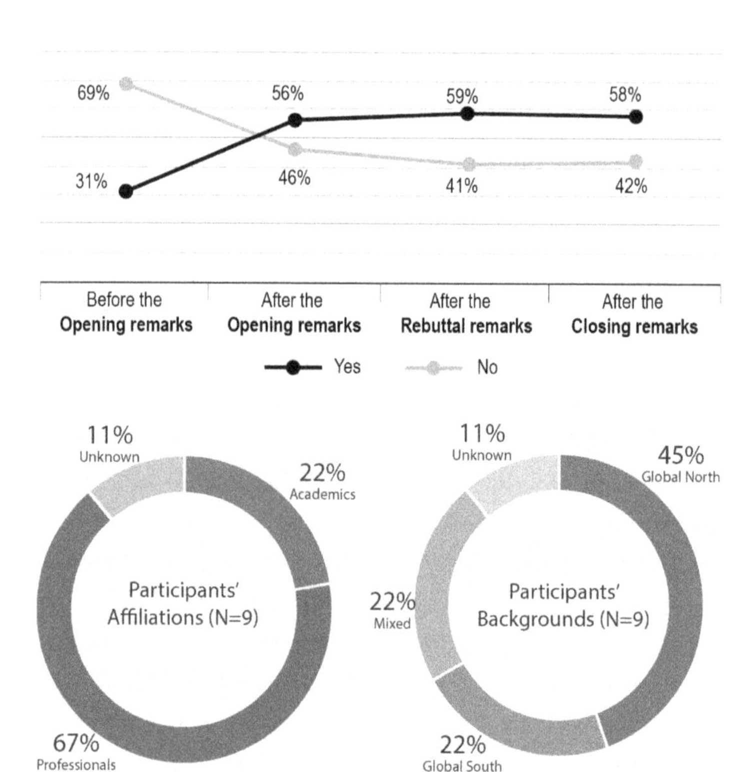

FIGURE 5.1 (*Top*) Results of votes at each stage of the debate. (*Bottom*) Distribution of affiliations and backgrounds among participants in the discussion.

1. Encampments, as an imperfect but acceptable immediate response to refugee crises, should be rethought and redesigned as more complete environments providing job opportunities and social services; this evolution might occur by involving urban planners and architects as well as other professionals capable of providing services and expertise beyond housing.

2. When refugees settle in urban and rural areas, they need to be well supported to avoid marginalization, exploitation, and stigmatization.

Participants raised many more questions that remain open-ended, such as issues of uncertainties and time management, the multiple definitions of a refugee, and the jurisdiction and responsibilities of different authorities in deciding how to shelter displaced people. These are questions that we must continue pondering collectively; the good news, as Crisp mentioned, is that international agencies dealing with refugees are aware of their importance and urgency.

NOTES

1. Marlen de la Chaux, Helen Haugh, and Royston Greenwood, "Organizing Refugee Camps: 'Respected Space' and 'Listening Posts,'" *Academy of Management Discoveries* 4, no. 2 (2018): 155–179.
2. Marc-Antoine Perouse de Montclos and Peter Mwangi Kagwanja, "Refugee Camps or Cities? The Socio-Economic Dynamics of the Dadaab and Kakuma Camps in Northern Kenya," *Journal of Refugee Studies* 13, no. 2 (2000): 205–222.
3. Elizabeth Holzer, "What Happens to Law in a Refugee Camp?," *Law & Society Review* 47, no. 4 (2013): 837–872.
4. Jamal Alnsour and Julia Meaton, "Housing Conditions in Palestinian Refugee Camps, Jordan," *Cities* 36 (2014): 65–73.
5. Shamima Akhter and Kyoko Kusakabe, "Gender-Based Violence Among Documented Rohingya Refugees in Bangladesh," *Indian Journal of Gender Studies* 21, no. 2 (2014): 225–246.
6. Jeff Crisp, "Forms and Sources of Violence in Kenya's Refugee Camps," *Refugee Survey Quarterly* 19, no. 1 (2000): 54–70; Japheth Nkiriyehe Kwiringira et al., "Experiences of Gender Based Violence Among

Refugee Populations in Uganda: Evidence from Four Refugee Camps," *Eastern Africa Social Science Research Review* 34, no. 1 (2018): 291–311; Jennifer Alix-Garcia et al., "Do Refugee Camps Help or Hurt Hosts? The Case of Kakuma, Kenya," *Journal of Development Economics* 130 (2018): 66–83.

7. Guglielmo Verdirame, "Human Rights and Refugees: The Case of Kenya," *Journal of Refugee Studies* 12, no. 1 (1999): 54–77; UNHCR, *UNHCR Policy on Alternatives to Camps* (Geneva: UNHCR, 2014).
8. According to UNHCR, more than five thousand dead migrants in the Mediterranean Sea, in 2016 only.
9. Elkanah Absalom et al., "Participatory Methods and Approaches: Sharing Our Concerns and Looking to the Future," *PLA Notes* 22, no. 5 (1995): 5–10.
10. See UNHCR Global Appeal 2015: Lebanon, https://www.unhcr.org/en-au/5461e607b.pdf; Francesca Ippolito, Gianluca Borzoni, and Federico Casolari, eds., *Bilateral Relations in the Mediterranean: Prospects for Migration Issues* (Cheltenham, UK: Edward Elgar, 2020).
11. In Lebanon, one of the countries of the world with the highest ratio of refugees versus native population, a lot has happened since our debate back in 2017. Briefly, the country's economic, political, and social systems have entirely collapsed, and Lebanon is now considered a failed state by several international organizations. Everlasting Middle East instability, superpowers' international conflicts, structural political weakness, endemic corruption, the disastrous explosion of the port in August 2020, and, above all, the unbearable burden of having to provide for the needs of a foreign population of refugees larger than one-third of the local one, all entangled, have led the country to this catastrophic situation on a scale comparable to a major disaster. In less than three years, the devaluation of the local currency reached unprecedented proportions, going from one dollar = 1,500 Lebanese liras (in September 2019) to one dollar = 140,000 Lebanese liras (in March 2022), leaving a huge chunk of the Lebanese population living on less than one dollar per day, in fact much poorer than the refugees, who continue to benefit from international aid from the UNHCR and other NGOs. What happened today is exactly what we feared would happen in our debate of 2017.

6

ON PARTICIPATION

Is Public Participation the Key to Success for Urban Projects and Initiatives Aimed at Disaster Risk Reduction?

WITH CHRISTOPHER BRYANT AND CAMILLO BOANO

While some specialists and decision-makers consider public participation a means to achieve effective and sustainable urban change, others see it as an end in itself. The former group typically argues that public participation adds value to decision-making by enabling open discussion concerning individual interests and expectations while also reinforcing the common good and safety, thereby achieving more impactful social justice.[1] They also argue that public participation allows decision-makers and experts to understand problems better and find more appropriate solutions for risk reduction and other issues of public interest.[2] Experts often emphasize the importance of "participating," seeing involvement as an independent outcome that increases social capital and resilience, strengthens a sense of community, and empowers community members and civic groups.[3]

Critics of public participation question both its effectiveness and its implementation.[4] They claim that participatory decision-making is often expensive and lengthy.[5] They note that elites and well-organized social groups often shape public opinion and take advantage of participatory activities to manipulate nonexperts and citizens.[6] Public participation, for instance, can enable those opposed to policies, programs, and projects to block them—the well-known colloquial term for one such form of opposition is NIMBYism, an acronym for "(I would support this project, but) Not In My Back Yard"—including, in some cases, measures disliked by some individuals but required by the greater public for disaster risk reduction.[7] Critics of public participation further contend that public influence over urban choices often produces solutions that are too costly or technically infeasible to implement, that cannot realistically be materialized, and that leave citizens cynical, frustrated, and distrustful of political and participatory processes.[8] Finally, participation too often confers an impression of public voice, inclusivity, and democracy on processes better characterized as tokenistic, manipulative, or deceitful.[9] In this sense, the unequal distribution of power among participants hampers the effectiveness of participation both as a means and as a goal.[10]

CHRISTOPHER BRYANT: IN DEFENSE OF PUBLIC PARTICIPATION

Public participation can be applied to any of the four sectors of urban planning—land use planning; environmental planning; developmental planning of economic, social, and cultural activities; and disaster risk reduction—that together form the strategic orientations of a community or territory. My response will focus first on urban land use, environmental, and development planning before tackling disaster risk reduction. Circumstances vary among territories, cities, and countries regarding how public participation is viewed and practiced; sometimes it is hardly practiced at all because of the lack of awareness of elected officials and planning professionals. However, public participation is increasingly viewed by its supporters as an essential element of managing change in a sustainable society.

What are the most important contributions of public participation? I identify four key ways public participation can benefit planning. First, public participation can integrate local knowledge into different kinds of planning. Second, it can involve recognizing that some local knowledge is more important than that of elected leaders and planning professionals; participants comprise people from all backgrounds and experiences, many of whom have valuable planning knowledge. Third, participation can draw upon the concern citizens have about what happens in their communities—where they work and live, and where their children grow up and may reside—and their resulting interest in helping to shape those communities. Finally, participation can tap into citizens' capacity to develop and manage initiatives.

Thus, public participation has already established its role in many municipalities as one of the keys to successful strategic development planning. However, to be effective, participation also requires efforts to mobilize and communicate with diverse populations, including young people, regarding their different and equally legitimate interests.

How does participation relate to projects and initiatives aimed at disaster risk reduction? Of course, people help their fellow citizens when a natural or human-based hazard strikes. But how can public participation contribute to disaster risk reduction efforts in the long term? I suggest integrating considerations of disasters and disaster risk reduction into long-term strategic development planning. This aim should also focus on building resilience and community solidarity.

Communities or territories need not have experienced a disaster to address public participation in disaster risk reduction initiatives; the probability of certain hazards, such as floods, tidal surges, forest fires, and earthquakes, can be appreciated by observing their impacts on similar communities in the region. Strategic development planning can address disaster risk reduction in several ways, such as improving protective infrastructure for coastal communities, preventing development in vulnerable areas, and preparing citizens to help others. Urban projects and initiatives aimed at reducing risks must be integrated into strategic development planning. Public participation in these initiatives is essential for several reasons. Local and regional politicians and professionals cannot be expected to have all the knowledge necessary to develop community resilience and solidarity. Citizens, with their local knowledge and commitments, can make substantial contributions to these causes. By participating in disaster risk reduction, citizens can help identify the most at-risk segments of the population and propose initiatives that can be independently managed. One participation initiative, for example, might involve training young students to assist other students in the event of a disaster, a solution put in place in the south of France, where many coastal communities have experienced devastating floods.

CAMILLO BOANO: CHALLENGING PUBLIC PARTICIPATION

Participation is important, but it rarely works. It is essential in any dimension of social life, including city-making, but rarely results in real emancipation. Despite improvements in the management of urban projects and urban design, Robert Chambers, Bill Cooke, Uma Kothar, and especially Erik Swyngedouw argue that "public urban participatory schemes" are "spaces where citizenship is constrained rather than activated."[11] Andrea Cornwall observes that invited and invented spaces of participation shape city-making.[12] In externally established and expert-managed invited spaces, urban citizens are invited to participate individually or collectively in social organizations. In contrast, invented spaces are grassroots-led forms of collective political mobilization. Invited space typically characterizes disaster risk reduction initiatives; engagement is often imposed, and participation is circumscribed by preestablished norms. Simply, my answer to the moderator's opening question is no, participation is not the key to success for urban projects and initiatives aimed at disaster risk reduction.

We must rethink the terms of participation by asking the following questions: What is a public? Who initiates urban projects, and who benefits from their implementation? Who is asked to participate? Who can participate?

To address these questions, we must situate participation within political discourses among the forces, structures, problems, and strategies that shape our collective and entangled lives. State, municipal, or agency-led participation have become normalized components of city governance and programming, resulting in a standard definition of participation as a platform for sharing information and listening to citizens' voices. Having invited residents to participate in various incentives, states, municipalities, and agencies often use participants' actions to claim their mandate to implement their solutions. Such invited spaces are complicit as they cement the positions of elites, prevent residents from taking ownership of participatory processes, and undermine the voices of marginal or vulnerable groups. These shortcomings box participation into a set of best practices and actions and reduce its emancipatory potential.

As a political concept, participation sheds light on how collective actions can change outcomes at any scale. Participation enables residents to actively—if imperfectly and contingently—demonstrate their rights as urban citizens. Considering competing practices of city-making, I believe that we must resist the depoliticization of participation. To use Holston's terminology, spaces of participation—demarcating citizens who can act and perform their rights, ambitions, and obligations—are inherently insurgent.[13]

I argue that participation is not a given component of urban policies and processes but is instead a practiced, claimed, and demonstrated action. Andrea Cornwall and John Gaventa contend that urban participatory mechanisms offer a space for citizens to become active "makers and shapers" in determining their mode of citizenship, rather than passive "users and choosers" of assigned participatory spaces that often perpetuate their exclusion though apparent means of inclusion.[14] According to Jacques Rancière, the role of politics is to problematize the "natural" order of things and verify equality.[15] Spaces in which participants can exercise their citizenship to challenge structural power dynamics are radical, transformative political arenas.

States, municipalities, and agencies often employ participation as a technique for city management in ways that remain largely ineffective for disaster risk reduction. As a complex mediation of space, power, and knowledge, straddling urbanization, capitalism, and democratic processes, participation impels its practitioners to question their expertise and control.

Without actively interrogating practices of recognition, liberation, and activation, as well as the definitions of "common" and "use," participation seldom works in a long-lasting way. This is especially the case in recovery and disaster risk reduction processes stressing trauma and resilience, according to which difficulties and complexities of post-disaster realities compromise participants' visions for their futures.

CHRISTOPHER BRYANT'S REBUTTAL REMARKS

It is certainly important to rethink the terms of participation. Who initiates urban projects and programs in the territory? What do we mean by urban projects? And how can we link experiences of public participation in urban planning and urban projects with disaster risk reduction initiatives?

How public participation is organized, who organizes it, and who must be mobilized and consulted to encourage citizens to participate are important factors to consider in assessments of public participation. Public participation will not be the same in all spaces and locations. In many cases, public participation is also a process that will evolve.

While municipal authorities can take the lead in initiating the process of public participation in, for example, a process of strategic development planning by and for the citizens, it is important to ensure that ultimately the citizens take over the process, create the vision of the community, and discuss what strategic orientations are necessary to achieve the vision. These strategic orientations can include disaster risk reduction.

Proper leadership is essential for successful public participation. Public participation requires leaders who honestly desire to share the processes of planning and preparing for the future with citizens and who accept, participate in, or even manage the citizen-run initiatives that are integral to the development of communities. This has recently occurred with greater frequency and with more substantial success.

I have been involved in several initiatives in which towns used a professional consultant to develop a strategic development plan. Unfortunately, the consultants only consulted elected officials and the towns' professional planners. Little was implemented of the final plans. As a last resort, the towns decided to undertake a strategic planning process by and for citizens, which gave rise to many initiatives to be implemented and managed by teams of citizens; these were realized and completely successful.

One of the mayors presented the process his community had initiated at a major municipal conference. When an audience member asked him how he felt about citizens taking and managing initiatives, the mayor replied that when he witnessed the citizens' group discussions, he realized that his most important role was to support citizen-led and -designed initiatives. He saw these initiatives as convergent rather than opposed to civic objectives of the municipality.

The ability of leaders to communicate effectively with the different segments of interest in the community is crucial; this approach involves understanding how people communicate with one another and encouraging their participation in planning and reflection processes. Real public participation does not focus on elites. Teenagers, for instance, play important roles in most communities because they represent their future. Therefore, public participation must be prepared carefully. Sometimes public participation requires addressing the values and perceptions of elected officials and municipal planners with the intent of getting them to critically assess their own values and biases around public inputs. One certainty is that public participation can achieve what otherwise would not be undertaken—what would otherwise be impossible-to-realize civic objectives—including initiatives in which mobilized citizens consider and manage initiatives to reduce vulnerability to disasters.

Engaging citizens and encouraging their involvement in planning and initiatives is essential to the functioning of a democratic society. As Amal Mohammed Jamal commented, citizen participation is the best way to move toward sustainable reconstruction or constructive urban change.

PARTICIPANT'S COMMENTS

Kristen MacAskill, Assistant Professor, Department of Engineering, University of Cambridge

Participation is not the key to success for initiatives aimed at disaster risk reduction, but that is not to say that I believe participation should not happen or is disadvantageous; indeed, it is important that engineers and planners work with communities where and when appropriate. Planning and executing a project include complexity and challenges. Various critical factors determine a project's success, including funding mechanisms, well-structured legislation, collaboration across organizations, and the role of experienced practitioners in conceptualizing, designing, and delivering the project. Additionally, trust plays a role in determining the success of a project. Participation is less critical if those involved have trust in decision makers. Of course, this trust must also be deserved.

CAMILLO BOANO'S REBUTTAL REMARKS

I argued that it is impossible to be against participation, but it is crucial to interrogate how the practice of participation is not simply symbolic or a cover-up. To enact change, participation must be transformative. It should be implemented as part of a radical political project through which participants can exercise their citizenship to challenge existing structural contexts. Disaster risk reduction is related to the risks and vulnerabilities faced by social groups. An isolated technical and instrumental approach to public participation must be avoided.

Local epistemologies, methodologies, and networks are fundamental to understanding risks, identifying hazards, and exploring needed interventions. When community members are involved in decision-making processes, they become more aware and engaged in the design and planning of adaptation. Community participation across all phases of the design and development of safer houses can increase mutual understanding among stakeholders, enable knowledge sharing, provide suitable options for specific contexts, and empower users. These processes of bottom-up engagement and involvement must be based on the experiences, knowledge, and capacities of locals rather than those of master planners. Innovative practices advance ways of producing space and knowledge that originate in local people and communities. In such cases, their collective endeavors are central to developing resilient and adaptive approaches to housing and city-making. Specifically, urban poor groups and other grassroots organizations are fundamental to the production of the whole city; they keep the city going. Their role in city-building sparks political engagement and contributes to a sense of ownership over the process and its results.

Somsook Boonyabancha, the secretary general of the Asian Coalition of Housing Rights, reminds us that "instead of the city being a vertical unit of control, these smaller units—people-based and local—can be a system of self-control for a more creative, more meaningful development."[16] The capacity to be an active participant and exercise agency is key to Boonyabancha's claim. Carole Pateman states that "participation, as far as the majority is concerned, is participation in the choice of the decision makers," suggesting that full participation emerges when each individual member of a decision-making body has equal power in determining the decision.[17] Such equality of input is certainly both an ideal and a trajectory, but one that seems impossible to achieve; such full participation depends on a shared level of knowledge and of specific forms of communication within codes and norms unfamiliar to many people. Jeremy Till refers to this ideal as "transformative participation," as it is based on the recognition of "the political aspects of space, of the vagaries of the lives of users, of different modes of communication and representation, of an expanded definition of architectural knowledge and of the inescapable contingency of practice."[18] In Jacques Rancière's terms, this "inclusion of the excluded" is a flawed approach to thinking politically about public participation and democracy.[19]

The logic of identification is ineffective, as it will reproduce risks and vulnerabilities rather than effect transformative changes. Participation in urban design and city-making, as well as in disaster risk reduction, requires abandoning the belief in expert knowledge in favor of willing acceptance of diverse forms of knowledge. Participation manifests only when plans, projects, and interventions are shaped by the multiple voices of insiders in a cacophony of imperfections. To this end, Markus Miessen's book with the prophetic title *The Nightmare of Participation*, though not providing us with a recipe for success, prompts us to ponder the three positions through which modes of proactive participation can become meaningful: "attitude, relevance, responsibility."[20]

CHRISTOPHER BRYANT'S CLOSING REMARKS

In some jurisdictions, the direct participation of the public, if included at all, is not very well developed. It is a shame for governments in these jurisdictions. Public participation can introduce significant change and innovative projects, including those linked to disaster risk reduction. Such jurisdictions must work with citizens by involving them in small groups to discuss specific issues and move forward based on their insights. A collaborative process between governments and participants is important because we cannot only rely on elected officials and professionals to make decisions on collective matters. Elected officials do not necessarily represent all community interests. They may think they do, only to realize that they do not; the planners and elected officials then resort, in many cases, to hiring external consultants to aid in governance and management.

The challenge is not just to recognize real issues. Of equal import is ensuring that the decisions made are reasonable and contribute to broad community objectives that the citizens have approved and appreciate. This is true whether the government is a central actor or not. There are many resilience- and community solidarity–building initiatives that do not rely on significant government involvement. Our primary challenge in public participation is not only recognizing real civic issues but also ensuring that reasonable decisions that can contribute to the community, city, and territory are made.

To pursue this focus, we must recognize that the different actors across various levels of government are people and citizens. If we consider the interests of actors in planning and managing development, we note that mayors, elected officials, deputies elected to parliaments, and other decision-makers have varied interests apart from their affiliated constituents. This analysis formed part of one of the first conceptual frameworks I constructed in the early 1990s, based on more than twenty-five years of research into the dynamics of territorial development—a framework that has been used by researchers interested in the dynamics of social change to analyze many different cultures. These leaders' diverse interests include implementing community objectives as well as personal goals such as generating income, personal development, helping friends achieve their objectives, and thwarting the plans of rivals.

Any analysis of the dynamics of change in territories, cities, and towns must also consider the informal networks in which decision-makers function. For instance, I was once invited to attend a meeting of a Table de Concertation (a deliberative civil society group) involving approximately twenty organizations from the focus county. A committee member drove me to the meeting and, during our trip, received a call from a contact. The caller sought to pressure my colleague to ensure that certain interests and options were discussed and presented at the meeting. This may seem nothing too dramatic. However, when elected officials fail to put aside their personal interests and allow undue influence or financial incentives to affect their decisions, they may be prosecuted and even imprisoned.

Although one hopes that elected officials and planning professionals will put their personal interests well behind those of their constituent communities, a way to avoid the dilemma is to have significant public participation. Engaging the public in long-term discussions and decision-making, including initiatives and plans to deal with disaster risk reduction, can redress potential power imbalances.

CAMILLO BOANO'S CLOSING REMARKS

I have focused, in my short interventions, on the participation aspect of this debate, hoping not to suggest that participation is unnecessary but rather to instill, as several commentators reiterated, the need to link it with emancipation, resistance, and recognition.

Participation is power. Cities are developed from the complex intersections of space, power, and knowledge. Any action that attempts to shape the city—in the context of this debate, with a disaster risk reduction emphasis—must navigate entanglements among spaces (material and otherwise); knowledge (expert, everyday, or, to borrow from Till, from "citizen experts"); and power. Any actions carried out by planners, architects, or citymakers involve some sort of complicity with existing power dynamics. Even if participants proceed with the best intentions, certain issues always arise during participatory processes, making them always imperfect. But when participants acknowledge the existence of power structures and deal with them responsibly, humbly, and with a sense of resistance, such processes may at least be more honest, acceptable, and potentially emancipatory.

Let me close with a point about the urban issues of participation. The built environment is a social product and political assemblage in which a multiplicity of institutions are forged and continue to shape society and be shaped as politics evolve. Cities are socially constructed environments in which humans transform and control nature, capital, and knowledge to produce urban forms of habitation. Cities emerge from the interactions among different complex practices, which are influenced by regimes of power and interface with the actions of urban poor and marginalized communities in shaping everyday urbanism. Any urban project and condition must come to terms with how its social complexity is constructed, produced, and imagined by various dynamics of power.

Edgar Pieterse refers to such urban realities as a "constitutive complexity" that needs to be mapped out not simply by listing the different powers at play in an urban context but also by highlighting "the practices through which power operates, the symbolic and the material effects power produces and its performance."[21] His approach requires us to think broadly about the forms and agencies involved in city-making, as well as the range of actors beyond experts participating "creatively, logistically, and politically" in the planning, development, and construction of cities.

Edgar Morin once postulated that embracing complex thinking does not demand abandoning a cause-effect logic (more participation, better effect) or adopting agonistic dialectics (formal/informal, order/disorder, visible/invisible, legal/illegal, planned/unplanned, etc.).[22] Complex thinking about urban systems recognizes that products and effects—such as urban form, density, movement of residents, rent increases, risks, and vulnerabilities—at the same time produce and are produced, are intended and unintended, designed and not designed, inherited and new. When knowledge about a socially relevant problem remains uncertain or disputed, and when there is a great deal at stake for those concerned by such problems, complex thinking has to operate effectively in a transdisciplinary manner. Ultimately, participation processes are responsible for creatively and incessantly engaging people in new, collaborative forms of spatial production. To paraphrase Italian philosopher Giorgio Agamben when speaking about the transdisciplinary intensity of philosophy, we can say that participation is not a fixed discipline but rather "an intensity that can suddenly give life to any field: art, religion, economics, poetry, passion, love, even boredom."[23]

PARTICIPANT'S COMMENTS

Amal Mohammed Hassan Jamal, Department of Architecture and Urban Planning, Benghazi (Garyounis) University, Libya

Based on my field study and documented heritage research on the Old Town of Ghat in Libya, the historical capital of the Kel Azjer Tuareg's former sultanate, I can confirm that the community's collective efforts and direct involvement in decision-making helped rebuild and restore the town after it was twice hit by natural disasters—a mountain rockslide and a rare heavy Saharan rain.

Despite the fact that this oasis town was formerly under the firm control of Tuareg leaders, the different ethnic groups that inhabited its built environment, comprising Arabs, Africans, and a Tuareg majority, had a say in how and where to rebuild their town. This ancient structure has survived because of the direct and collective involvement of those participating in decision-making, which led to the creation of a more homogenous social and built structure.

Public participation is not difficult or impossible. I am familiar with cases in which broadcasting public meetings enabled people to reach united decisions that represented and reflected the opinions of diverse participants. Muting or omitting public participation will lead to the failure of urban projects and disaster risk reduction initiatives.

MODERATOR'S CLOSING REMARKS: PUBLIC PARTICIPATION IS STRONGLY LINKED TO SUCCESS IN DISASTER RISK REDUCTION, DESPITE DOUBTS ABOUT ITS IMPLEMENTATION

This debate addressed the role of public participation in the success of urban projects and initiatives aimed at disaster risk reduction. Given the widespread acceptance of the public participation paradigm, it is not surprising that most participants defended positive outcomes of participation and the value of participating. About 70 percent of voters consistently defended this position over the twenty days of the debate (figure 6.1). Although some participants identified drawbacks, limits, and secondary effects of public participation, their observations had a negligible impact on the general perception of the effectiveness and value of public participation. Moreover, most participants seemed to find an intrinsic value in public engagement that is not necessarily dependent on its contextual conditions or specific characteristics.

Participants consider public participation as an opportunity for citizen involvement in decision-making processes, one that creates a balance in power relations among citizens, authorities, and experts. (See, for example, the work by Wisner and by Gaillard and Maceda.)[24] Fewer than 30 percent of participants raised doubts about the intrinsic benefits of public participation. For

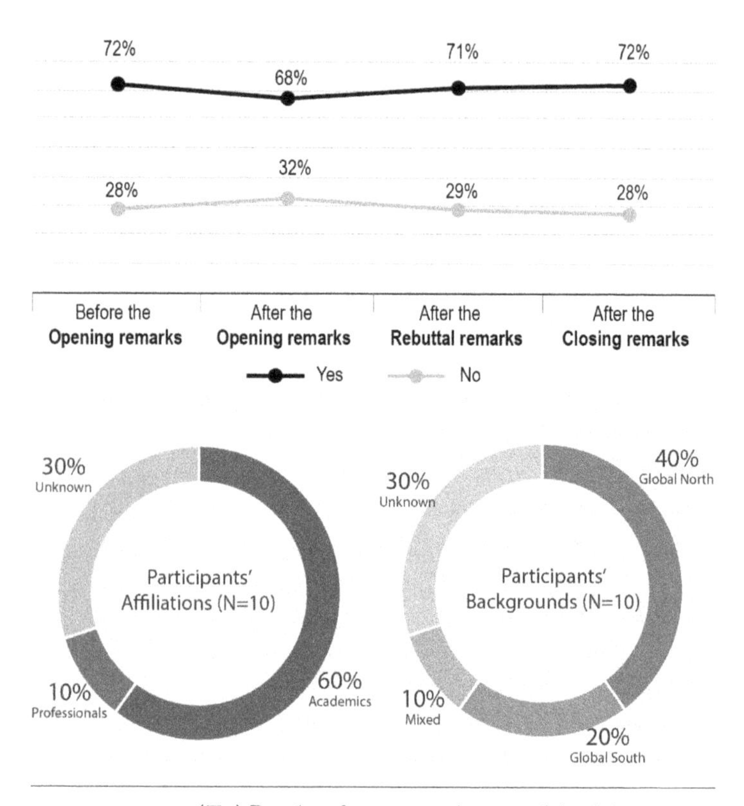

FIGURE 6.1 (*Top*) Results of votes at each stage of the debate. (*Bottom*) Distribution of affiliations and backgrounds among participants in the discussion.

this minority, the benefits of public participation depend on significant variables, notably the form of power involved; they preferred "claimed and demonstrated" participation over the "given" type of participation. This exchange raised an important question: Are the problems of public participation accidents of implementation and thus mistakes that could be corrected and avoided? Or are they the consequence of the concept of "making/letting people participate"?

From the supporting perspective, the lack of "mobilization and communication with the different segments of legitimate interests in the community or territory" threatens participatory decision-making processes. Bryant points to authorities' tendency to make decisions based on their interests, neglecting the view of all legitimate stakeholders (he fails to clarify, however, how they obtain this legitimacy). He contends that sharing "the planning process with citizens" and involving citizens' capacities for constructing and managing local initiatives are common challenges for leaders such as mayors and supporting professionals. Even so, leaders and stakeholders largely determine the outcomes of an initiative.

From the opposing perspective, Boano points to existing literature that distinguishes between invented and invited spaces of participation. He recognizes that invited forms of participation, which are established and managed by often external experts, typically "cement elites' position, prevent resident ownership of processes, and undermine voices of marginal or vulnerable groups." He sees increased value in "grassroots-led forms of collective political mobilization," which shape transformative spaces where participants exercise their citizenship to challenge structural contexts. Finally, Boano recognizes the existence of sophisticated power structures that complexify and make participatory decision-making processes "always imperfect."

Bloggers and participants developed nuanced arguments about public participation. They raised interesting points about the intrinsic and extrinsic value of public participation, the links between public participation and the political nature of urban planning, and the relationship between public participation and freedom to change. Participants were also concerned with the ethical parameters of public participation, the reasons for failure in public participation, those responsible for fixing its flaws and limits, and issues of planning and participation fatigue. These results confirm that public participation is an unavoidable, and important, aspect of disaster risk reduction and urban interventions. The results also invite us to continue questioning its ethical value, whether intrinsic or extrinsic, and its role in distributing power among authorities, experts, and lay citizens.

NOTES

1. Ben Wisner, "Self-Assessment of Coping Capacity: Participatory, Proactive and Qualitative Engagement of Communities in Their Own Risk Management," in *Measuring Vulnerability to Natural Hazards: Towards Disaster Resilient Societies*, ed. Jörn Birkmann (Tokyo: UN University Press, 2006), 316–328; Giacomo Rambaldi et al., "Resource Use, Development Planning, and Safeguarding Intangible Cultural Heritage: Lessons from Fiji Islands," *Participatory Learning and Action* 54, no. 1 (2006): 28–35.
2. Peter A. Kwaku Kyem, "Power, Participation, and Inflexible Institutions: An Examination of the Challenges to Community Empowerment in Participatory GIS Applications." *Cartographica: The International Journal for Geographic Information and Geovisualization* 38, no. 3–4 (2001): 5–17.
3. Stephen Connelly, "Looking Inside Public Involvement: How Is It Made So Ineffective and Can We Change This?," *Community Development Journal* 41, no. 1 (2006): 13–24.
4. Robert Chambers, "Reflections and Directions: A Personal Note," in *Participatory Learning and Action 50: Critical Reflections, Future Directions*, ed. H. Ashley (London: IIED, 2004), 23–34; Renee A. Irvin and John Stansbury, "Citizen Participation in Decision Making: Is It Worth the Effort?," *Public Administration Review* 64, no. 1 (2004): 55–65.
5. Sarah Bradshaw, "Engendering Development and Disasters," *Disasters* 39, Suppl 1 (2015): S54–S75.

6. Divya Chandrasekhar, "Digging Deeper: Participation and Non-Participation in Post-Disaster Community Recovery," *Community Development* 43, no. 5 (2012): 614–629.
7. Robert Chambers, *Whose Reality Counts? Putting the Last First* (London: Intermediate Technology, 1997); Bradshaw, "Engendering Development and Disasters."
8. Connelly, "Looking Inside Public Involvement."
9. Sherry R. Arnstein, "A Ladder of Citizen Participation," *Journal of the American Planning Association*, no. 35 (1969): 216–224.
10. Giacomo Rambaldi et al., "Practical Ethics for PGIS Practitioners, Facilitators, Technology Intermediaries and Researchers," *Participatory Learning and Action* 54, no. 1 (2006): 106–113; John Forester, "Challenges of Deliberation and Participation," in *Reading in Planning Theory*, ed. Susan S. Fainstein and Scott Campbell (Malden, MA: Wiley-Blackwell, 2012), 206–213.
11. Chambers, *Whose Reality Counts?*; Bill Cooke and Uma Kothari, *Participation: The New Tyranny?* (London: Zed, 2001); Uma Kothari, "Power, Knowledge and Social Control in Participatory Development," *Participation: The New Tyranny* (2001): 139–152; Erik Swyngedouw, "Governance Innovation and the Citizen: The Janus Face of Governance-Beyond-the-State," *Urban Studies* 42, no. 11 (2005): 1991–2006.
12. Andrea Cornwall, "Introduction: New Democratic Spaces? The Politics and Dynamics of Institutionalised Participation," *IDS Bulletin* 48, no. 1A (2017): 1–10.
13. James Holston, *Insurgent Citizenship: Disjunctions of Democracy and Modernity in Brazil* (Princeton, NJ: Princeton University Press, 2009).
14. Andrea Cornwall and John Gaventa, "From Users and Choosers to Makers and Shapers: Repositioning Participation in Social Policy," *IDS Bulletin* 31, no. 4 (2000): 50–62.
15. Jacques Rancière, *Disagreement: Politics and Philosophy* (Minneapolis: University of Minnesota Press, 1999).
16. Somsook Boonyabancha, "Unlocking People Energy," *Our Planet: The Magazine of the United Nations Environment Programme* 16, no. 1 (2005): 22–23, 23.
17. Carole Pateman, *Participation and Democratic Theory* (Cambridge: Cambridge University Press, 1970), 13.

18. Jeremy Till, "The Negotiation of Hope," in *Architecture and Participation*, ed. Peter Blundell-Jones, Doina Petrescu, and Jeremy Till (London: Spon, 2005), 19–41.
19. Jacques Rancière, "From Politics to Aesthetics?," *Paragraph* 28, no. 1 (2005): 13–25.
20. Markus Miessen, *The Nightmare of Participation* (Berlin: Sternberg, 2010).
21. Edgar Pieterse, *City Futures: Confronting the Crisis of Urban Development.* (London: Zed, 2013).
22. Edgar Morin, "Restricted Complexity, General Complexity," in *Worldviews, Science and Us: Philosophy and Complexity*, ed. Carlos Gershenson, Diederik Aerts, and Bruce Edmonds (Hackensack, NJ: World Scientific, 2007), 1–25.
23. Antonio Gnolio, "Philosophy as Interdisciplinary Intensity – An Interview with Giorgio Agamben," *Religious Theory*, 2017, https://jcrt.org/religioustheory/2017/02/06/philosophy-as-interdisciplinary-intensity-an-interview-with-giorgio-agamben-antonio-gnolioido-govrin/.
24. Ben Wisner et al., *At Risk: Natural Hazards, People's Vulnerability and Disasters*, 2nd ed. (London: Routledge, 2004); Jean Christophe Gaillard and Emmanuel A. Maceda, "Participatory Three-Dimensional Mapping for Disaster Risk Reduction," in *Participatory Learning and Action: Community-Based Adaptation to Climate Change*, ed. Holly Ashley, Nicole Kenton, and Angela Milligan (Nottingham, UK: Russell, 2009), 109–118.

7

ON AID

Does Aid Actually Aid in Avoiding Disasters and Rebuilding After Disasters?

WITH ANNA KONOTCHICK AND JASON VON MEDING

For decades, scholars and think tanks have debated the effectiveness of aid in reducing poverty. This debate builds on previous arguments about the effectiveness and value of aid, focusing on its role in disaster risk reduction and post-disaster reconstruction and recovery. Scholars, think tanks, celebrities, and politicians have claimed aid is crucial to preventing famines, diseases, and death. They argue that donors' money can be used to solve basic problems, whether sanitation, vaccination, education, housing, or infrastructure.[1] Aid can also be used to fund monitoring activities and learn from interventions. More importantly, supporters of aid contend that traditional markets alone cannot resolve housing and infrastructure deficits. They note that the poor are often stuck in what economists call "poverty traps";[2] that is, millions of people are poor precisely because they live in poverty. Slum dwellers, for instance, find it difficult to escape poverty because they pay proportionally more for services and infrastructure than wealthier citizens.[3] Foreign aid is needed to break these vicious economic cycles and replace them with more resilient cycles that provide opportunities for the most vulnerable.[4] For defenders of aid, lack of funds, rather than existing mechanisms or structures of aid, poses the greatest problem in reducing poverty. In their view, criticisms of aid merely provide an excuse for not donating money.

Conversely, critics of aid believe that too much money is wasted on aid. They hold that initiatives seldom produce positive long-term change and, in many cases, create more damaging than desirable results. They contend that donors' money is spent on Band-Aid solutions that lack enduring impact.[5] Furthermore, they argue that aid distribution is dictated by political agendas, replicates forms of neocolonialism, focuses too much on technology transfer, creates dependencies, or bypasses formal governments and authorities.[6] Finally, critics assert that aid is often based on centralized schemes produced by overconfident and idealistic decision-makers with little knowledge of what is needed "on the ground." For them, ideology drives aid, which, as a result, lacks the performance incentives and accountability mechanisms found in competitive—and typically more efficient—markets.[7]

ANNA KONOTCHICK: AID AIDS IN AVOIDING DISASTERS AND REBUILDING AFTER DISASTERS

For disaster-affected communities, aid can reduce vulnerability to natural hazards and improve the recovery process. However, systems currently in place to deliver aid are imperfect, and humanitarians, experts, and activists need to work within institutions—NGOs, governments, academic institutions, the private sector, or community groups—to make aid more effective, fair, and just.

Research proves that investments in disaster preparation pay off financially: one dollar invested in disaster mitigation saves society about six dollars.[8] Those savings are predominantly directed to average families. In countries like the Philippines, where 40 percent of the population has experienced homes damaged by natural disasters, hazard-resistant shelters save financial resources and ensure stability and security in times of crisis.[9] Early warning systems, evacuation plans, and cadres of professionals and community volunteers trained in first aid save lives daily.[10] Every life is precious.

In addition, post-disaster recovery efforts, from individual response efforts to global agreements, are becoming more effective and responsive. Greater emphasis on direct cash transfers places decision-making in the hands of disaster-affected families; cash transfers simultaneously reinforce local livelihoods and reduce previous inefficiencies in aid provision by NGOs or implementing agencies.[11] The Grand Bargain, an agreement between donors and aid providers, requires that more effort and resources be spent on localizing financial, technical, and coordination resources in a transparent manner.[12] The World Humanitarian Summit reiterated that aid should not only be the project of traditional aid players like NGOs, but requires coalitions of local communities, governments, and the private sector to effect change. The "Communication as Aid" initiative focuses on empowering disaster-affected voices in recovery efforts and decision-making.[13] Systems ensuring humanitarian accountability to affected and vulnerable populations improve every year.[14] While such efforts save lives and restore stability, they require resources beyond those of standard aid efforts.[15]

More must be done to ensure good stewardship of aid resources. Aid can and must evolve. Paradigm shifts are necessary to address the root causes of vulnerability to natural hazards. We know that structural issues and power inequalities result in natural disasters having a disproportionate impact on the poor.[16] This fact has spurred important social movements like climate justice.[17] The poor occupy lands that flood more often. They have less political power to demand disaster mitigation infrastructure. Ethical humanitarian practitioners must recognize this inequality. We must recognize how our interventions can either disrupt or be complicit with unjust systems. We must move beyond thinking of simplistic solutions or projects that narrowly focus only on the symptoms of vulnerability.

The question posed by the moderators shifts the notion that any one agent, particularly foreign experts, can solve poverty and vulnerability to natural hazards. A new generation of humble double agents is required to lessen the impact of natural disasters on the poor; such double agents must dare to recognize their complicity in the political and economic systems that cause poverty and vulnerability to natural hazards, yet remain motivated to transform those systems.[18] Aid is a collective challenge and responsibility, regardless of whether one works for an NGO or a community group or in the government, private sector, or academia.

JASON VON MEDING: AID DOES NOT ASSIST IN AVOIDING DISASTERS AND REBUILDING AFTER THEM

I begin by speaking about the gross historical injustice perpetrated against the Global South by the North. Since 1492, Europe relied on the exploitation of those it could subjugate for its development and enrichment. Without recognizing this context, any conversation about whether humanitarian aid works would be lacking. Today, the dominant development model is also one that celebrates social "progress" from a Eurocentric and paternalistic perspective. Countries that have been pillaged are called poor, underdeveloped, or developing. Beset by the labels that conceal crimes, Michael Perenti calls our attention to the set of social relations that has been forcefully imposed on countries.[19]

Let us not make a false assumption that poverty is a first stage of cultural development or naturally occurs in society. Inequality and injustice underpin systems of governance, diplomacy, and trade. I cannot deconstruct the concept of the "poverty trap" in this response, but I consider Jeffrey Sachs's "more aid" and William Easterly's "no aid" approaches as different ways of solving the wrong problem.[20] That is, both approaches operate according to status quo assumptions about poverty, inequality, and development.

I argue that humanitarian aid does not really help in avoiding disasters or in rebuilding after they occur. If we are to avoid disasters, we must aim at reducing risk and stopping risk creation. An ahistorical approach to disaster risk and development, such as that espoused by Bill Gates or Steven Pinker, simply kicks the can down the road and assures us that everyone can be a winner. I believe instead that the overprivileged, the oppressive, and the elite must sacrifice.

Most aid money spent relative to disasters focuses on recovery; additional spending on prevention is essential. Yet only a small fraction of aid money goes to frontline local organizations.[21] Profiteers move rapidly into post-disaster scenarios, often under the excuse of humanitarian aid. Although humanitarian agencies are staffed by inspiring and genuine people, it remains a sector steeped in the paternalistic and neocolonial discourse of charity.

At best, charity provides a Band-Aid solution; at worst, it obscures a status quo that perpetuates structural violence to maximize profit. Most disturbingly, some of the most enthusiastic proponents of philanthropy continue to benefit handsomely from an exploitative system. What about fighting the system itself, for which humanitarian aid and development practitioners were once recognized? The sector must get back to challenging the status quo rather than assenting to tyranny.

PARTICIPANT'S COMMENTS

Thomas Johnson, Associate Lecturer, School of Architecture and Built Environment, University of Newcastle, Australia

I am currently collecting data in Bangladesh on how the humanitarian sector is reducing disaster vulnerability during the Rohingya refugee crisis. Like the point raised by Vanicka on Bhaktapur, the government in Rohingya has rejected large donations that did not align with its agenda. All foreign-funded projects require FD7, an approval process with the NGO Affairs Bureau of Bangladesh, and require that humanitarian projects cannot last longer than six months (previously three months).[a] In this context, it seems that the humanitarian sector could reduce the long-term vulnerability of displaced people if the government did not enforce such restrictions and if they had continuous funds to help develop the host community. However, this is simply impossible.

Other responses highlight short-term hazard interventions rather than the root causes of vulnerability. I do not think decision-makers do so from a lack of understanding of disaster vulnerability, but rather as a consequence of funding limitations and measures of effectiveness. Donations usually peak within the first year of a response and dwindle from that point, resulting in Band-Aid fixes that do not reduce long-term disaster vulnerability. Some interesting approaches, such as anticipatory finance, provide compelling alternatives to donations.

[a] Abdul Kadir Khan and Tiina Kontinen, "Impediments to Localization Agenda: Humanitarian Space in the Rohingya Response in Bangladesh," *Journal of International Humanitarian Action* 7, no. 1 (2022): 14; Caitlin Wake and John Bryant, "Capacity and Complementarity in the Rohingya Response in Bangladesh," ODI Global, December 5, 2018, https://odi.org/en/publications/capacity-and-complementarity-in-the-rohingya-response-in-bangladesh/.

ANNA KONOTCHICK'S REBUTTAL REMARKS

The arc of the moral universe is long, but it bends toward justice.

—Dr. Martin Luther King

Von Meding and I agree that vulnerability to natural hazards is a political problem that demands political solutions.[22] Progress toward more just systems and aid practices is not a given; it must be hard-won. Like past social movements, progress in aid requires engaging aid, political, and economic systems at all levels and requires compromised institutions to hire double agents sympathetic to change.[23] Systemic political changes must be envisioned as part of our humanitarian goal beyond immediate lifesaving efforts. But how can and do humanitarians put such a goal into practice?

First, we can learn from the Global South. Muhammad Yunus of Bangladesh has revolutionized the formal banking sector and international development practice, as well as international development practice, in Bangladesh and globally.[24] Slum/Shack Dwellers International's "Know Your City" campaign inspired post-earthquake Haiti's enumeration and neighborhood recovery efforts.[25] The homegrown development agency and microfinance institution BRAC is one of the largest humanitarian responders to the Rohingya crisis in Cox's Bazar; nearly half a million people are served by their tube wells.[26] Humanitarian responders are increasingly forming a transnational community.

Second, we must recognize that unjust political and economic systems exist across different geographies; they exist within nations and communities and not only along the simplistic dichotomy of Global North and South. Disasters overwhelm local systems and institutions and can thus be considered as moments when systems and institutions are vulnerable to external influence, progressive or otherwise. As humanitarian workers, we pursue a mandate to work directly with marginalized communities in disaster-affected regions. As such, we have the opportunity and agency to represent their needs when they are unable to do so and to challenge unjust local and global systems. Vested political and economic interests fill any available opening or suggestion in humanitarian discourse, exacerbating existing inequalities.[27] Humanitarians raise awareness of these needs and call for action and change through research and monitoring, placing resources in the hands of vulnerable peoples and communities, or engaging in quiet diplomacy.[28] Their ability to raise awareness is also why repressive governments try to limit their mobility or power.[29]

Finally, we can continue to self-critique and learn. The humanitarian aid community recognizes that more has to be done to improve. We continue to hold systems accountable to affected communities, establish clear codes of conduct and financial reporting, and ensure affected populations are included in decision-making.[30] Experts and practitioners are increasingly realizing that aid operates in political contexts, while new tools like political economy assessments are becoming standard practices to help design practitioners' actions.[31]

Disasters are the time when vulnerable communities need humanitarian organizations to listen, magnify their voices, and open spaces to negotiate their recovery. The status quo will not do this for them.

JASON VON MEDING'S REBUTTAL REMARKS

Konotchick has provided many excellent examples of the attempts by well-meaning practitioners and organizations to use aid as a vehicle for disaster recovery and risk reduction. These achievements are fantastic—and somewhat miraculous—given that little aid reaches frontline communities. The system is dysfunctional primarily because of its intrinsic link to global capital.

I am fond of the article referenced by Konotchick, "Praxis in the Time of Empire." Ananya Roy states that "this article is written with the hope that praxis in the time of empire can turn the heart of power into a profound edge of struggle and dissent." If we apply this aim to the current debate on aid in disaster recovery, risk reduction, and development, we can, by proxy, make a case for the importance of double agents in humanitarian practices. As my opening remarks suggested, humanitarian practitioners used to be known for their ability to work both sides of policy agendas.

But in the twenty-first century, who are these double agents, and what do they seek to change? Do they seek to change the "imperfect systems of aid delivery," as Konotchick suggests? In my opinion, focusing on how aid can "work better" assumes the continuation of the status quo as it shapes the economy, development, inequality, and poverty. I would be delighted if the humanitarian sector staged the grounds for struggle and dissent, but most agencies are so concerned with ensuring continued funding that they cannot fathom structural change. Will they ever bite the hand that feeds? We must also not forget that we are in a global mass extinction event. We are pushing toward limits that threaten human society, if not our survival as a species. Aid is a key component of a development engine that has brought us to this point. Is it ethical to help maintain such a destructive system?

Although aid is supposed to be the altruistic arm of development, for every dollar of aid that developing countries receive, they lose twenty-four dollars in net outflows.[32] Overall, the overwhelming exploitation of developing countries is not helping the cause of justice. Arguably, charity exploits human compassion to prop up a failing model of aid—is it being wasted? Daniel Raventós and Julie Wark argue that charity is not a gift and that the impossibility of reciprocating charity ensures the perpetuation of class structures.[33]

I am intrigued by the idea that we can be both complicit and subversive in humanitarian work. I recall similar assertions from Wendell Berry and Noam Chomsky, which Ananya Roy refers to as an "ethics of doubleness."[34] She argues that such an approach is suitable "under conditions of extreme power where the ethical autonomy required to articulate disavowal and refusal might be lacking." I wonder if the ethics of doubleness applies to humanitarian aid practitioners in the context of disaster recovery and risk reduction. Can one operate as an effective double agent, and if so, what is the end goal of such a practice?

PARTICIPANT'S COMMENTS

Ekatherina Zhukova, Senior Lecturer, Department of Political Science, Lund University, Sweden

My wording to describe participation—"transforming"—is somewhere between "replacing," to describe von Meding's approach, and "adjusting," to describe Konotchick's. To replace means to dismantle and create something new from scratch, an ideal solution with which I strongly sympathize in theory but find difficult to implement in practice. To adjust is to build on and improve an existing system. It is easier to implement this approach in practice, but it does not solve the problem of inequalities. I would continue searching for alternative solutions like South-South cooperation to bring more practice into von Meding's approach and integrate the voice of aid recipients to complicate Konotchick's approach.

ANNA KONOTCHICK'S CLOSING REMARKS

We cannot accept that working within a compromised system is not good enough. Working within current systems is sometimes necessary, especially when there are no viable alternatives. Even if we do not work specifically in the aid sector, we all are somehow complicit in the current system. We pay taxes to government systems that perpetuate uneven trade deals and cripple poor economies.

I quoted Dr. Martin Luther King Jr. in my rebuttal because he inspires those willing to fight for change, even when the obstacles to change seem insurmountable. He envisioned a more just society but did not deconstruct U.S. government institutions that had systematically oppressed people of color. He revised them, made them evolve, and worked to introduce the Voting Rights Act, but much work remained to be done.[35] Paradigm shifts start small. As humanitarians, we work with institutions—perhaps our governments, or those where we work—to make them more responsive to the needs and rights of marginalized and vulnerable people.

An anecdote about Canaan inspires my optimism. Canaan is now the third largest city in Haiti. Its homes, schools, and five hundred kilometers of roads have been built since the 2010 earthquake, but not by the government, aid, or the private sector. In a display of unbelievable unity amid tragedy, sixty thousand earthquake-affected families invested more than one hundred million dollars of personal resources in new homes and infrastructure, building their vision of a more just and livable city. However, their vision for a hopeful future is anything but assured.[36] In asserting people's rights to the city, Canaan is a battleground.[37] People faced evictions, and their homes were bulldozed to rubble.[38] Government provided no aid, water network, paved roads, or public schools. Canaan also endured the greatest cholera outbreak in the country.

Through humanitarian aid, Canaan has now acquired its first social services and infrastructure. This aid is not charity, but responds to the civic rights of the city's residents. Accurate geospatial mapping and population estimates, generated by aid agencies with the collaboration of the community, changed government budget allocations. The Ministry of Health vaccination rates plummeted when Canaan's population was included, so humanitarian organizations developed new outreach programs. Formerly gray, empty splotches on city maps transformed into detailed neighborhoods.[39] More needs to be done to improve the lives of residents, but change has started. Such a progressive shift in policy, one oriented toward upgrading and extending services to informal neighborhoods, is never assured. Disasters can easily introduce opportune moments for governments or wealthy actors to clear the poor from the land.[40]

Humanitarians can foster progressive steps because of their direct connection to affected communities and their mandate to serve the most vulnerable. Double agents in governments will always be influenced to serve those in power, so they also depend on humanitarians to balance public discourse and help them in representing the most vulnerable.[41]

JASON VON MEDING'S CLOSING REMARKS

Does aid work? I believe that alongside its somewhat limited achievements, it regrettably props up a socioeconomic status quo—one structurally based on scarcity thinking—that causes misery for billions of people. This is why I argue against the proposition of the debate.

Of course, many worthwhile programs in disaster recovery and risk reduction have been undertaken under the auspices of aid. I am not concerned if good is being done, but whether the root causes that prompt a need for aid are being addressed. If not, is aid just a Band-Aid response to inequality, poverty, and disaster?

Michael Perenti asks, "Why has poverty deepened while foreign aid and loans and investments have grown?"[42] As he argues in *The Face of Imperialism*, investments, loans, and aid work exactly as intended, against the interests of the communities they claim to serve. He claims that the ultimate purpose of the aid sector is to "serve the interests of global capital accumulation."

I agree with Konotchick's argument that "systemic political changes must be seen as part of our humanitarian goal" but wonder how many charitable organizations and large donors align with such a mandate. Aid is arguably a tool for wealthy individuals, corporations, and countries to appear virtuous without systemic change. Some philanthropists admit that charity helps preserve wealth—this is not justice.

For example, Bill Gates could lobby for less strict patent laws and allow millions of people in the Global South to access generic medicines. Instead, he chooses to lobby aggressively for laws that protect his sources of wealth accumulation, in turn perpetuating global health crises. Then, without recognizing the irony of his decisions, he provides aid to ameliorate the situation he helped cause.

An analysis of the microfinance model of aid reveals that, despite broad adoption and powerful anecdotes, the long-term impacts of aid are not so positive. Milford Bateman and Ha-Joon Chang argue that "continued support for microfinance in international development policy circles cannot be divorced from its supreme serviceability to the neoliberal/globalization agenda."[43] Critics argue that the model is rooted in "the myth of the heroic individual entrepreneur, the rags to riches fairy tales."[44] The same claim might be made regarding the rhetoric and practice across aid: aid is patently compatible with the status quo.

There is no doubt that many organizations operating within neoliberal systems are doing some good for some people. Should they simply cease providing aid? Of course not. Konotchick and I agree that double agents are important to bring about change, but we must be honest about the scale of change required to achieve a society in which provisions for basic needs are rights rather than privileges. Tinkering at the edges of the status quo will not benefit those targeted by humanitarian aid, the most oppressed and marginalized individuals and communities.

PARTICIPANT'S COMMENTS

Vanicka Arora, PhD candidate, Institute for Culture and Society, Western Sydney University, Australia

A consideration of Bhaktapur, Nepal forwards a recent example regarding the agency and participation of aid recipients. After protracted discussions with the government of Germany, the local municipality of Bhaktapur rejected aid amounting to thirty million euros for post-disaster reconstruction.[a] Multiple players with opposing agendas that could not be mediated were involved in making this decision. The need of the local municipality to assert its financial and technical independence from foreign influence and national intervention formed one agenda. Another agenda involved the growing pressure from community groups to source local materials and technologies. A third agenda concerned ethnic politics within Bhaktapur that had begun fomenting before the earthquake. These entwined decision-making processes suggest that aid recipients are not always passive and without a voice or agency—they are non-monolithic and nonbinary actors.

While the North-South, South-South arguments have been brought up repeatedly in this debate, aid is not simply a transaction among nation-states, transnational organizations, and governments, but moves through multiple levels of bureaucracy, politics, and inefficient systems. While it is important to consider international politics and neocolonial agendas in public participation, the importance of local and national politics cannot be discounted.

[a] See "Clash of Cultures in Bhaktapur," *Nepali Times*, June 2018, https://www.nepalitimes.com/banner/clash-of-cultures-in-bhaktapur.

THE MODERATOR'S CLOSING REMARKS: THE ETHICS OF AID— REPLACING THE AID SYSTEM OR CHANGING IT FROM WITHIN?

Aid faces several problems. Our two panelists and almost all participants in our debate agree that aid needs a paradigm shift, but they do not agree on how to reach this goal. Anna Konotchick, a senior officer in the field of housing and urban development, invites aid workers to recognize how their interventions might disrupt or inadvertently be complicit with injustice. For Konotchick, institutions can have benevolent and transformative "double agents" who support vulnerable people while promoting the development of more just social, economic, and political systems. Jason von Meding, a scholar on a crusade to halt the impacts of colonialism and disaster risk creation, argues that complying with the current system of aid is no longer an option. According to von Meding, oppressive systems demand more radical changes. His view was supported by 57 percent of voters, but it raises many questions.

Given the environmental challenges we face, should aid be stopped? Can we provide aid to help the most vulnerable? Can

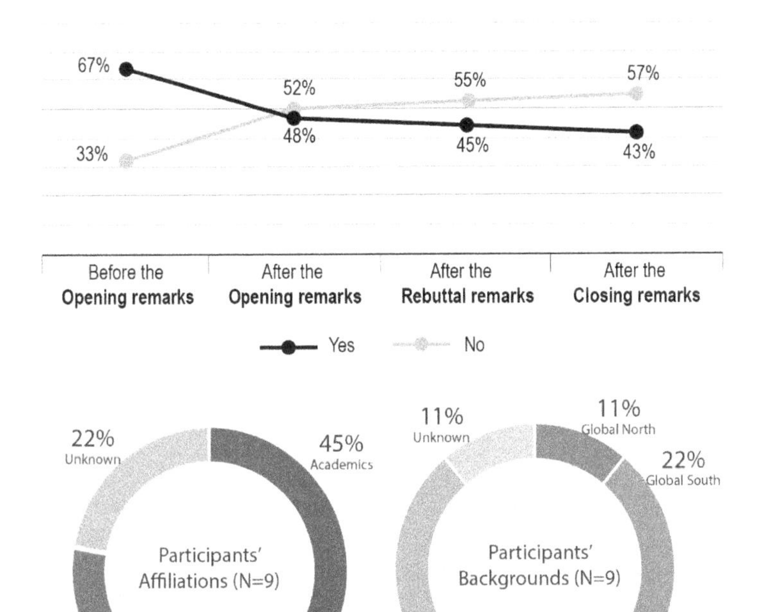

FIGURE 7.1 (*Top*) Results of votes at each stage of the debate. (*Bottom*) Distribution of affiliations and backgrounds among participants in the discussion.

poor countries and communities in the Global South—known to be the most vulnerable to environmental hazards—cope independently with the effects of climate change, mass displacement, and poverty?

Von Meding and other scholars and activists cited in this debate have argued that aid reproduces forms of colonialism, imperialism, and domination. In this sense, aid mechanisms are too embedded in neoliberal and capitalist practices to be fixed. Yet these same critics often refer to data obtained or published by agencies or think tanks funded by aid. They also claim that the voices of the most vulnerable are often unheard in capitalist discourses and practices, neoliberal policy, and geopolitics. In emphasizing such points, they fail to acknowledge fully that charities and aid organizations sometimes amplify the voices of the most neglected, oppressed, and marginalized, as raised by Konotchick in her closing remarks. Finally, they claim that aid often resorts to Band-Aid solutions that fail to produce structural, positive changes and tackle the root causes of vulnerability, but stop short of suggesting alternatives. If the current aid system is to be replaced, as more than half of voters seem to agree, what type of mechanisms or solution should replace it?

Is South-South aid, for instance, a viable alternative to traditional North-South aid distribution dynamics? Respondents were uncertain, pointing to examples in which patterns of domination and exploitation are reproduced within countries of the Global South. Should governments avoid mingling with aid agendas? Again, participants remained uncertain, suggesting cases in which the deregulation of humanitarian aid permitted abuses and favored partisan agendas. Is local aid better than international aid? No. Local charities are typically constrained by political interests that hinder their capacity to produce radical change.

Konotchick advanced a clear answer to these questions, stating that attempting to fight the system is probably naïve, counterproductive, and overwhelming. Instead of getting rid of aid altogether, she argues for the role of aid double agents, who can be simultaneously complicit and subversive. Yet her approach also raises several questions. Given that aid organizations and charities depend on sustained funding, can they enforce structural changes while contesting the status quo that supports them? What are the moral implications of playing these double games?

What ethical boundaries must be drawn when we are complicit with dodgy governments or institutions? Is complicity a form of endorsement?

Both panelists agree that most aid workers are well-intentioned and ethical individuals but that the system in which they operate has many deficiencies. We do not believe participants are convinced that the aid industry behaves as a single, homogeneous system. Instead, given the multitude of stakeholders engaged in the aid delivery process and their divergent interests, aid increasingly appears as a constellation of multiple systems and networks behaving in dynamic and not necessarily overlapping ways. If multiple systems coexist, should we discard them all at once? Otherwise, which systems must be the target of subversion or the source of complicity? How can replacing, subverting, or complying with systems be achieved when we are dealing with moving targets and diverse contexts?

NOTES

1. Simon Feeny, "The Impact of Foreign Aid on Poverty and Human Well-Being in Papua New Guinea," *Asia Pacific Development Journal* 10, no. 2 (2003): 73–93; Nasim Shah Shirazi, Turkhan Ali Abdul Mannap, and Muhammad Ali, "Effectiveness of Foreign Aid and Human Development," *Pakistan Development Review* 48, no. 4 (2009): 853–862.
2. Aart Kraay and Claudio Raddatz, *Poverty Traps, Aid, and Growth* (Washington, DC: World Bank, 2005).
3. Muhammad Mizanur Rahaman and Tahmid Saif Ahmed, "Affordable Water Pricing for Slums Dwellers in Dhaka Metropolitan Area: The Case of Three Slums," *Journal of Water Resource Engineering and Management* 3, no. 1 (2016): 15–33.
4. Howard Pack and Janet Rothenberg Pack, "Foreign Aid and the Question of Fungibility," *Review of Economics and Statistics* 75, no. 2 (1993): 258–265.

5. Paul Collier and David Dollar, "Aid Allocation and Poverty Reduction," *European Economic Review* 46, no. 8 (2002): 1475–1500.
6. Alberto Alesina and David Dollar, "Who Gives Foreign Aid to Whom and Why?," *Journal of Economic Growth* 5, no. 1 (2000): 33–63.
7. Craig Burnside and David Dollar, "Aid, Policies, and Growth," *American Economic Review* 90, no. 4 (2000): 847–868; William Easterly and Tobias Pfutze, "Where Does the Money Go? Best and Worst Practices in Foreign Aid," *Journal of Economic Perspectives* 22, no. 2 (2008): 29–52.
8. National Institute of Building Sciences, "National Institute of Building Sciences Issues New Report on the Value of Mitigation," January 11, 2018, https://www.nibs.org/news/national-institute-building-sciences-issues-new-report-value-mitigation.
9. Vincenzo Bollettino et al., *Perceptions of Disaster Resilience and Preparedness in the Philippines*, Harvard Humanitarian Initiative, June 2018, https://hhi.harvard.edu/publications/perceptions-disaster-resilience-and-preparedness-philippines.
10. European Commission Humanitarian Aid, *Preparing for Disaster Saves Lives*, 2007, https://ec.europa.eu/echo/files/media/publications/dipecho_en.pdf.
11. ODI Global, "10 Things You Should Know About Cash Transfers," Accessed December 24, 2022, https://www.youtube.com/watch?v=Vo8DZytvjXg.
12. Agenda for Humanity, "Initiative: Grand Bargain," https://agendaforhumanity.org/initiatives/3861.
13. CDAC Network, https://www.cdacnetwork.org.
14. "Core Humanitarian Standard on Quality and Accountability," updated 2024, https://corehumanitarianstandard.org.
15. UN News, "Global Aid Appeal Targets More Than 93 Million Most in Need Next Year," December 4, 2018, https://news.un.org/en/story/2018/12/1027521.
16. Joyce Coffee, "Climate Disasters Hurt the Poor the Most: Here's What We Can Do About It," Governing, February 13, 2018. https://www.governing.com/commentary/col-disasters-disadvantaged-climate-justice.html; Mary B. Anderson, "The Impacts of Natural Disasters on the Poor: A Background Note," 2001https://www.gfdrr.org/sites/default/files/publication/The%20Impacts%20of%20Natural%20Disasters%20on%20the%20Poor.pdf.

17. Patrick Bond, *Politics of Climate Change Justice: Paralysis Above, Movement Below* (Scottsville, South Africa: University of Kwazulu-Natal Press, 2012).
18. Ananya Roy, "Praxis in the Time of Empire," *Planning Theory* 5, no. 1 (2006): 7–29.
19. Michael Parenti, *The Face of Imperialism* (New York: Taylor & Francis, 2016).
20. Jeffrey Sachs, "The Case for Aid," *Foreign Policy*, January 21, 2014; William Easterly, "Can Foreign Aid Buy Growth?," *Journal of Economic Perspectives* 17, no. 3 (2003): 23–48.
21. IRIN, "Where Is All the Money Going? The Humanitarian Economy," n.d., http://newirin.irinnews.org/the-humanitarian-economy/.
22. Jason Hickel, *The Divide: A Brief Guide to Global Inequality and Its Solutions* (London: Penguin, 2017).
23. Ananya Roy, "Praxis in the Time of Empire," *Planning Theory* 5, no. 1 (2006): 7–29.
24. Muhammad Yunus, *Banker to the Poor: Micro-Lending and the Battle Against World Poverty* (New York: Public Affairs, 2007); Marguerite S. Robinson, *The Microfinance Revolution: Sustainable Finance for the Poor* (Washington, DC: World Bank, 2001).
25. Slum Dwellers International (SDI), "The Know Your City Campaign," 2023, http://knowyourcity.info/.
26. "BRAC's Humanitarian Response in Cox's Bazar: Strategy for 2018, April 30, 2018, https://reliefweb.int/report/bangladesh/brac-s-humanitarian-response-cox-s-bazar-strategy-2018.
27. Naomi Klein, *The Shock Doctrine: The Rise of Disaster Capitalism* (New York: Holt, 2007).
28. Amnesty International, *The State of the World's Human Rights* (London: Amnesty International, 2018).
29. Timothy McLaughlin, "Burma Considers Law That Could Restrict Work of United Nations, Nongovernmental Groups," *Washington Post*, March 19, 2018, https://pulitzercenter.org/reporting/burma-considers-law-could-restrict-work-united-nations-nongovernmental-groups.
30. "Core Humanitarian Standard on Quality and Accountability"; IFRC, "Code of Conduct for the International Red Cross and Red Crescent Movement and NGOs in Disaster Relief," January 1, 1994, https://www.ifrc.org/document/code-conduct-international-red-cross-and-red

-crescent-movement-and-ngos-disaster-relief; International Aid Transparency Initiative (IATI), "Connecting IATI and Un Ocha's Financial Tracking Service to Streamline Humanitarian Reporting," May 17, 2018, https://iatistandard.org/en/news/connecting-iati-and-un-ochas-financial-tracking-service-to-streamline-humanitarian-reporting/.

31. Diana Cammack, "Draft Working Document: USAID Applied Political Economy Analysis (PEA) Field Guide," February 1, 2016, https://www.usaid.gov/sites/default/files/documents/2496/Applied%20PEA%20Field%20Guide%20and%20Framework%20Working%20Document%20041516.pdf.
32. Jason Hickel, *The Divide: A Brief Guide to Global Inequality and Its Solutions* (London: Penguin, 2017).
33. Daniel Raventós and Julie Wark, *Against Charity: How Capitalism Loves Humanity* (Chico, CA: AK Press, 2018).
34. Ananya Roy, "Praxis in the Time of Empire," *Planning Theory* 5, no. 1 (2006): 7–29, 23.
35. History, "Voting Rights Act of 1965," updated 2023, https://www.history.com/topics/black-history/voting-rights-act.
36. Jacob Kushner, "An Ungoverned City," *U.S. News & World Report*, January 15, 2019, https://www.usnews.com/news/cities/articles/2019-01-15/canaan-haiti-is-a-test-of-how-a-city-can-self-govern.
37. David Harvey, "The Right to the City," *City Reader* 6, no. 1 (2008): 23–40.
38. Reliefweb, "Thousands More at Risk of Forced Eviction," February 4, 2014, https://reliefweb.int/report/haiti/thousands-more-risk-forced-eviction.
39. Janet Byrne, ed., *Know Your City: Slum Dwellers Count* (Cape Town, South Africa: SDI, 2018).
40. Khristine Alvarez, "Resilient City Making: Dispossession and Urbanisation in Post-Ondoy Manila," paper presented at the Urban and Regional Development Annual Conference, Mexico City, July 2016; Maike Grabowski and Anne Ritter, "Political Changes – Changes in Human Rights Policies?," *Observer* 3, no. 1 (2011).
41. James A. Paul, "NGOs and Global Policy-Making," Global Policy Forum, June 2000, https://archive.globalpolicy.org/component/content

/article/177-un/31611-ngos-and-global-policy-making.html; Véronique de Geoffroy and Alain Robyns, "Strategies Used by International NGOs to Influence Public Policy," March 30, 2010, https://www.urd.org/en/review-hem/strategies-used-by-international-ngos-to-influence-public-policy/?artpage=2-4.

42. Michael Parenti, *The Face of Imperialism* (New York: Taylor & Francis, 2016), 56.
43. Milford Bateman and Ha-Joon Chang, "Microfinance and the Illusion of Development: From Hubris to Nemesis in Thirty Years," *World Economic Review*, no. 1 (2012): 13–36.
44. Madeleine Buntin, "Is Microfinance a Neoliberal Fairytale?," *Guardian*, March 9, 2011, https://www.theguardian.com/global-development/poverty-matters/2011/mar/09/microfinance-neoliberal-fairytale.

PART III

WHAT OUGHT TO BE

Living with Risk or Relocating in Isabela de Sagua, Cuba

GONZALO LIZARRALDE, TAPAN DHAR, AND LISA BORNSTEIN

THE SLOW AGONY OF THE DROWNED

Vintage photos of Isabela de Sagua depict a village of timber houses built on stilts. Women and children elegantly dressed in white clothes ride boats that navigate through the water running under wooden verandas and porticos. As though in a tropical Venice, wood bridges appear like aerial sidewalks connecting homes and facilities. These passages extend to the land and seem to have served as the village's main public meeting areas.

The villagers depicted in the photos are Cubans from the early twentieth century. Isabela de Sagua, now a small village in the northern littoral of Cuba, was once a key Cuban seaport, active in sugar trading and well connected to other cities by train. Today, the village houses fewer than three thousand people and has lost its economic value and strategic commercial importance (see figure III.1).

The aerial sidewalks and port decks have disappeared, thousands of residents have left, the historic port facilities are almost

FIGURE III.1 A street in Isabela de Sagua.

gone, and very few structures on stilts remain standing. The village has been destroyed and rebuilt several times over the past hundred years. Whereas archival photographs illustrate its massive destruction by a tropical storm in 1933, most of the recent damages were caused by Hurricane Kate in 1985 and Hurricane Irma in 2017.

When our research team met with community members and leaders from Isabela de Sagua in 2011 and 2018, we were told a story of loss, but also one of pride and desire to move forward. Villagers explained how satisfied they were to live in this place, full of history and surrounded by an astonishing landscape. Their homes—vernacular one- and two-story units built of wood and painted in pastel colors—are small and simple, but they are well ventilated, have comfortable verandas, face the ocean, and are suited to fishing activities and other demands of Caribbean lifestyles.

Residents know that Isabela de Sagua is prone to seasonal hurricanes and tropical storms. Today, it is also at risk of droughts as well as floods and erosion because of climate-change-induced

FIGURE III.2 The Soviet-style residential project developed in Nueva Isabela.

sea level rise.[1] But local leaders explained to us how they have learned to live close to the sea.[2] Locals know how to evacuate, protect their belongings and animals when storms are approaching, and rebuild and maintain the timber structures when the rain ends.[3] They fear that if they leave their homes or accept relocation, they might face the same fate as those who moved to Nueva Isabela.

Nueva Isabela is a government initiative created in response to the destruction caused by Hurricane Kate. Following the disaster, the government in La Havana wanted to relocate hundreds of families from Isabela de Sagua to a safer place, identifying a piece of land about eleven kilometers from the coast where they built several prefabricated apartment blocks (see figure III.2). The four-story buildings, which many locals refer to as Soviet-inspired architecture, rise on empty land disconnected from community services and far from Sagua La Grande, the main city in the area. Planning did not include public and recreational spaces, and few facilities were built in the new development. Villagers who moved to the prefabricated apartments feel disconnected from the ocean, and many have suffered from depression and other mental health problems. Whereas some decided to return to Isabela de Sagua, others kept their apartments and continue to commute to their home village. Fishers who moved to the new units have lost their livelihoods and miss their spiritual connection to the sea.

In general, locals believe that Nueva Isabela provides an inadequate solution to ongoing risks. While trying to reduce the exposure to hurricanes and tropical storms, the new village created new vulnerabilities among relocated residents. Both local scholars and residents agree that the new development lacks the beauty and character of the historic village. The project was also a missed opportunity to rehabilitate and highlight the heritage and culture of the old town.

Recently, international politics and a new hurricane collided to extend Nueva Isabela a bit more. Located in the path of hurricanes and storms, weakened for decades by political isolation and an absurd embargo imposed by American authorities, and dependent on resources and imports from abroad, Cuba is a particularly vulnerable island state. However, the Cuban socialist government takes climate adaptation, public housing, and disaster risk reduction seriously.[4] Recent policy in the country bans new construction and the reconstruction of houses and public facilities in flood-prone areas, including those threatened by sea level rise.[5] Yet the government and residents cannot afford the constant rebuilding and repairs needed after hurricanes, floods, and tropical storms.[6] Several plans have been advanced to build hurricane-resistant public housing in safer areas, mostly involving the deployment of prefabricated solutions in new residential developments.[7]

After Hurricane Irma, the friendly socialist government of Venezuela donated about fifty prefabricated houses to be built in Nueva Isabela. Given that the donations were made alongside oil contributions and international aid offered by former Venezuelan president Hugo Chavez, the units were nicknamed *petro casas* (petro-houses).[8] Much like the Soviet-style apartment blocks, the new single-story *petro casas* are imported on the land without incorporating public space, community facilities, or parks. Featuring no trees or urban décor, the suburban-like open space between units is rarely used.

When our team visited Nueva Isabela, we found a place where residents obtained a disaster-resistant shelter with running water, sewage, and electricity. But the new town and homes are a rather sad solution, ill-adapted to the culture, livelihoods, and lifestyles of Cuban villagers.[9] Most Cubans share a perception that, in making such housing decisions, the government was acting in good faith to reduce risk exposure and protect lives. Even so, the new town creates more problems than solutions.[10]

Regarding the fate of Isabela de Sagua, a Cuban blogger wrote about "the slow agony of the drowned," reflecting on the sorrow caused by repeated destruction, which inevitably leads to slowly fading places and cultures.[11] But in Isabella de Sagua, human solutions that have failed to consider the cultural and social expectations of people are responsible for loss.[12] When deciding "what ought to be," well-intentioned decision makers have created new problems, such as social inequalities, that only aggravate the damage caused by recurrent natural hazards.

Today, plans are in place in Cuba to relocate other villages to safer areas. Authorities fear that climate change impacts, including sea level rise, droughts, and the increasing intensity and frequency of meteorological events, will only worsen the conditions of people living in coastal areas. They claim that a lack of resources limits the systematic and continuous rebuilding and repair of houses, infrastructure, and public buildings.[13] Even so, hundreds of residents from other villages in the northern littoral, such as Carahatas, resist relocation and lobby for changes in policy so that they can continue living in their original locations.[14] These villagers, who are mostly fishers, aim to convince technocrats and politicians that they have the knowledge and skills required to live close to the ocean. They prefer continuity to other forms of "adaptation" to climate change.

The case of Isabela de Sagua, as well as Carahatas and other coastal villages in Cuba, exemplifies some of the tensions and difficult decisions that we identify in this book. These case studies illustrate how hard it has become to make decisions to mitigate the effects of global warming. They also remind us how difficult it is to design a built environment that mitigates future risk while responding to the current needs of households, as well as broader cultural rituals, traditions, meanings, and expectations.

The history of disasters and reconstruction is replete with mistakes made by authorities resulting from corruption, negligence, racism, colonialism, elitism, or a basic lack of empathy. But the Cuban case shows that mistakes are also made by authorities acting in good faith. The complexity of current problems makes it difficult for authorities to agree upon the right decisions for climate action, even if they are motivated by good intentions. Is relocation, for instance, the right decision to make when it becomes too expensive to rebuild and repair structures in disaster-prone locations? Can the reduction of risk exposure justify the creation of other forms of vulnerability? How do we proceed when a decision reduces one form of vulnerability but creates another?

In this book, we have tried to address the tensions and difficulties currently faced by decision- and policymakers interested in disaster risk reduction. In this last section, we address tough decisions related to international consultancy services; the implementation of compulsory regulations and voluntary performance indicators aimed at reducing risks; and the goal of adaptation to current and future risks, including those exacerbated by climate change effects.

When it comes to the built environment, deciding "what ought to be" is never easy. As global warming intensifies, urban planning, architecture, and landscape design must not only confront current challenges but also anticipate those that will emerge in the future. In times and places marked by limited resources and overstretched budgets, decisions must be made regarding what is good, just, sustainable, or necessary, as well as what is urgent and simply viable. To this end, decisions about the future demand a balance between conflicting objectives: what sacrifices are required today to be able to cope with risks tomorrow and to respond to the needs of future generations? What stakeholder claims are legitimate, and how can we anticipate how these might shift and change? How should we proceed when apparently good solutions contradict each other or create secondary effects?

As we shall see, decisions about how to plan and modify our current situation must face increasingly tough ethical dilemmas.

NEW MORAL DILEMMAS

Whereas economists and experts in international development have long debated the efficiency of external aid, they have not focused as much on the role of consultancy and knowledge transfer to cities.[15] For decades, defenders and opponents of international aid have clashed regarding the political pertinence and effective results of humanitarian workers who labor in harsh conditions "on the ground," often clad in the sleeveless khaki uniforms of NGOs and aid agencies.[16] However, the often invisible and increasingly powerful role of climate and resilience consultants and executives who meet in boardrooms, stay in expensive hotels, and sign the hundreds of policy briefs and recommendations that seek to dictate urban development, risk reduction, and climate action policy worldwide receives far less attention.

For some, international consulting has become a new form of colonialism. Defenders of the transfer of international expertise claim that the role played by consultants is key in coordinating global action, setting common goals, and making sure that clear objectives are identified and measured.[17] Yet the Isabela de Sagua case illustrates another ongoing reality in contemporary international aid. The *petro casas* and oil donated to Cuba by the Venezuelan government do not fit the common conception of rich countries "helping" poor ones in hard times, which more readily corresponds to an image of executives in Washington or Geneva wearing dark suits and deciding on resilience plans for a tropical city in the Caribbean. The former method of assistance reproduces some of the strategies and mistakes of politics between the Global North and South, while also disrupting standard notions of neocolonialism and economic investment that characterize American imperialism in Latin America or European assistance in Africa. Even so, urban and climate policy, including in Cuba, is increasingly influenced by the resilience and adaptation concepts, jargon, and metrics defined in boardrooms and fancy hotels in Northern countries.

THE DEBATES

On their own, cities often have limited resources and capacity to implement ambitious measures to reduce greenhouse gases or disaster risks. Several organizations, including the Rockefeller Foundation, Local Governments for Sustainability (ICLEI), and UN-Habitat, work with cities to support change. But is the presence of these organizations in urban governance morally neutral? Do they use their influence, status, and power in the right way? Our debate in chapter 8 addresses common issues surrounding climate and risk reduction consultancy work, while also revealing new trends in and controversies about geopolitics, economic power, and governance mechanisms. We situate this debate by considering the role of international agencies, consultants, and other agents often called "orchestrators" of change in reducing climate-related risks.[18] The debate leads us to explore the contradictions that emerge when international agencies and consultants try to orchestrate policy changes, along with the role that cities play in tackling global warming response.[19] Lorenzo Chelleri and Craig Johnson chart the advantages and disadvantages of having an increasing number of international organizations attempt to influence technocrats and local decision-makers toward adopting better climate plans. Are they succeeding in this effort? Do their actions address the effective reduction of vulnerabilities? Who benefits from the influence of these orchestrators, and what is the scope of that influence? The debate addresses these and similar questions to challenge the value and efficiency of climate and risk reduction reports and guidelines.

In chapter 9, we explore current debates regarding the imposition of regulations and standards to improve housing conditions in the Global South. Whereas houses in Isabela de Sagua fail to respond to most modern disaster-resistant standards, the *petro casas* and apartment buildings in Nueva Isabela resist earthquakes and hurricanes. Examining this Cuban case, some might wonder: should these standards justify the negative consequences of relocation? In this debate, Edmundo Werna and Brian Aldrich explore whether governments should devise and enforce standards for low-cost housing in the Global South. About 40 percent of people in the Global South live in informal urban settlements, where a lack of resources, services, and infrastructure puts millions in life-threatening living conditions.[20] Minimum standards, set by international agencies, are required to guarantee the safety of residents and the resistance of structures and infrastructure against hazards.[21] But climate change and new threats increasingly motivate professionals and regulators to upgrade current standards. How can places maintain a certain level of housing affordability while also improving thresholds of resistance and sustainability? In the case of countries in the Global South, where a strong informal sector undertakes the incremental construction of a large share of housing, the imposition and control of standards and codes become particularly challenging.[22] Some argue that elevating norms only motivates builders and users to resort to the informal construction sector. Other secondary effects of the implementation of standards include gentrification and a reduced interest of investors in low-income housing development. This debate demonstrates the need for a delicate equilibrium among standard compliance, affordability, and quality control, a balance that changes across different countries. Whereas Cuban authorities prioritize users' safety and try to keep informality at bay, authorities in other places allow the informal sector to prosper. The chapter concludes that there is a need to adapt construction practices to the reality of informal practices and housing markets in the Global South.

Whereas chapter 9 explores the definition and enforcement of compulsory standards, chapter 10 deals with the benefits and drawbacks of adopting voluntary ones. In this debate, Jared Blum and David Wachsmuth analyze the impact of green certifications, often developed in industrialized nations for developing countries. Can (and should) LEED, BREEAM, and other building and urban development standards be transferred from the North to the South? What are the consequences of such a transfer of construction and design practices? Are green certifications a productive way to reduce environmental damage in developing countries? This debate shows that standards might help achieve sustainable goals but that without proper adaptation to local conditions, they reproduce inequalities and fail to respond to users' real needs and expectations. An equilibrium between standardization and customization is required in housing, construction, and urban initiatives. Disregarding the specific considerations of territory, local culture, social rituals, common behaviors, traditional forms of construction, local materials, and the needs of users and communities increases vulnerabilities, erodes cultural values, and alienates residents. In some cases, compliance with international standards produces more negative effects than benefits.

In the final chapter, we address a theme discussed across several debates in this book: adaptation to climate change. Given the pace at which we are polluting the atmosphere and depleting ecosystems, is adapting to climate change our best choice moving forward? Should we prioritize adaptation, mitigation, or both? In this debate, we challenge the idea that adaptation is unavoidable. Deborah Harford and Silja Klepp explore the advantages and disadvantages of adopting a jargon of resilience in discussions of climate actions, particularly regarding the notion of adapting systems to current and future threats associated with global warming. What is obscured and what is revealed when we argue that people and their systems must "adapt"? What is implied when we assume that people have "adaptive capacities"? Harford and Klepp explore the power that the resilience narrative holds today, revealing both benefits and black spots of its adoption in policy, design, and aid projects. Although we ultimately do not have a choice about whether to adapt to existing climate realities, our way forward, as explored in this debate, depends on a more precise understanding of adaptation and its politics. Returning to the Cuban case, authorities, in adopting the narrative of climate adaptation, are promoting the relocation of villages identified as being in disaster-prone areas. Residents, however, prefer another form of adaptation, one that makes continuity and permanence possible, maintains their history and traditions, and prioritizes social cohesion and attachment to place. The case study of Cuba illustrates how the notion of adaptation can lead to different—even diverging—responses.

Part III brings us back to the issue of the explanations and narratives used to understand risk, disasters, and responses to them. Language is a concern cutting across almost all chapters of this book. Throughout the book, the debates show that how we define current and future threats significantly influences the way we respond to them. The definitions and concepts we mobilize shape behaviors and highlight specific aspects of reality while masking others. In this way, they allow us to understand the struggles of those facing vulnerabilities and social and environmental inequities, and shape the way we approach privilege and injustice. Language helps to frame the responses we adopt or impose upon others; the concepts and notions we employ help us analyze and understand experiences and lived realities. Chapters 9–11, building on others elsewhere in the book, reveal how the interpretations of vulnerabilities, risk, destruction, struggles, and actions contribute to our assessment of "what ought to be."

NOTES

1. Martha Thompson and Izaskun Gaviria, *Weathering the Storm: Lessons in Risk Reduction from Cuba* (Boston: Oxfam, 2004).
2. ADAPTO et al., *Reponses to Risk and Climate Change in Informal Settings in Latin America and the Caribbean: The Importance of Bottom-up Initiatives and Structured Dialogue.* (Montreal: Disaster Resilience and Sustainable Reconstruction Research Alliance—Oeuvre Durable, 2021).
3. David Smith et al., *Artefacts of Disaster Risk Reduction: Community-Based Responses to Climate Change in Latin America and the Caribbean* (Montreal: Université de Montréal, 2021).
4. Richard Stone, "Climate Adaptation: Cuba's 100-Year Plan for Climate Change," *Science* 359, no. 6372 (2018); Arturo Valladares, "The Community Architect Program: Implementing Participation-in-Design to Improve Housing Conditions in Cuba," *Habitat International* 38 (2013): 18–24.
5. Gonzalo Lizarralde et al., "A Systems Approach to Resilience in the Built Environment: The Case of Cuba," *Disasters* 39, Suppl 1 (2015): s76–s95.
6. Gonzalo Lizarralde, *Unnatural Disasters: Why Most Responses to Risk and Climate Change Fail but Some Succeed* (New York: Columbia University Press, 2021).

7. Gonzalo Lizarralde et al., "Does Climate Change Cause Disasters? How Citizens, Academics, and Leaders Explain Climate-Related Risk and Disasters in Latin America and the Caribbean," *International Journal of Disaster Risk Reduction* 58 (2021): 102173.
8. See details by the local press at https://www.granma.cu/cuba/2019-03-04/otra-isabela-se-levanta-04-03-2019-21-03-34.
9. Gonzalo Lizarralde, *The Invisible Houses: Rethinking and Designing Low-Cost Housing in Developing Countries* (London: Routledge, 2014).
10. Ernesto Aragón-Duran et al., "The Language of Risk and the Risk of Language: Mismatches in Risk Response in Cuban Coastal Villages," *International Journal of Disaster Risk Reduction* 50 (2020): 101712.
11. The text is available at https://cubaprofunda.wordpress.com/tag/isabela-nueva/.
12. Gonzalo Lizarralde et al., "We Said, They Said: The Politics of Conceptual Frameworks in Disasters and Climate Change in Colombia and Latin America," *International Journal of Disaster Prevention and Management* 29, no. 6 (2020).
13. Ministerio de Ciencia Tecnología y Medio Ambiente: CITMA, *Tarea vida: Plan de Estado para el enfrentamiento al cambio climático* 2020, March 14, 2020, https://www.ecured.cu/Tarea_Vida.
14. Rigel Hernandez, "Carahatas: La comunidad como pieza clave en la conservación del medio ambiente," *Flora y fauna: Ecología y Sociedad*, November 2014, 45–47; Aragón-Duran et al., "The Language of Risk and the Risk of Language."
15. Jakob Svensson, "Aid, Growth and Democracy," *Economics & Politics* 11, no. 3 (1999): 275–297; Dambisa Moyo, *Dead Aid: Why Aid Makes Things Worse and How There Is Another Way for Africa* (New York: Penguin, 2010); V. Desai, Kitchin Rob, and Thrift Nigel, "Aid," in *International Encyclopedia of Human Geography*, ed. Ron Kitchin and Nigel Thrift (Oxford: Elsevier, 2009), 84–90.
16. William Easterly, *The White Man's Burden: Why the West's Efforts to Aid the Rest Have Done So Much Ill and So Little Good* (New York: Penguin, 2006); Jeffrey Sachs, *The End of Poverty: How We Can Make It Happen in Our Lifetime* (New York: Penguin, 2005).
17. Ken Stern, *With Charity for All: Why Charities Are Failing and a Better Way to Give* (New York: Anchor, 2013).

18. Kenneth W. Abbott et al., *International Organizations as Orchestrators* (Cambridge: Cambridge University Press, 2015).
19. UN-Habitat, *Cities and Climate Changes— Global Report on Human Settlements 2011* (London: Earthscan, 2011); Jane Bicknell, *Adapting Cities to Climate Change* (London: Earthscan, 2009).
20. Mike Davis, *Planet of Slums* (London: Verso, 2006).
21. Stefan Greiving, Michio Ubaura, and Jaroslav Tešliar, eds., *Spatial Planning and Resilience Following Disasters: International and Comparative Perspectives* (Bristol, UK: Policy Press, 2016); Benjamin Herazo and Gonzalo Lizarralde, "The Influence of Green Building Certifications in Collaboration and Innovation Processes," *Construction Management and Economics* 33, no. 4 (2015): 279–298.
22. Edmundo Werna, "Shelter, Employment and the Informal City in the Context of the Present Economic Scene: Implications for Participatory Governance," *Habitat International* 25, no. 2 (2001): 209–227.

8

ON EXTERNAL AGENTS AND CONSULTANTS

Do International Agencies, Consultants, and Other "Orchestrators" Truly Help Cities Reduce Climate-Related Risks?

WITH LORENZO CHELLERI AND CRAIG JOHNSON

Cities are playing an increasingly active role in governing climate change action. Even in countries where national governments have done little to tackle global warming, municipalities are often seen as key to reducing both carbon emissions and vulnerabilities. But cities do not always have the resources or capacity to implement ambitious measures to reduce atmospheric pollution (mitigation) or disaster risk. In response, nongovernmental institutions are working with cities, rich and poor, to better tackle climate-related challenges. These organizations include 100 Resilient Cities, the Local Governments for Sustainability (ICLEI), C40, and UN-Habitat, as well as multiple city networks and international consulting firms.

Proponents of this approach contend that climate change and other risks must be addressed at a global scale by constructing international coalitions guided by common objectives. They argue that this approach makes it possible to combine funding and expertise from the public and private sectors. International consultants and agencies can fill gaps in municipal expertise and broaden participation to forge a more inclusive co-governance approach to global issues. Hybrid governance facilitates public awareness, reinforces relationships among cities, contributes to city-to-city and government-to-industry knowledge transfer, and offers a platform for promoting successful policy experiments and capacity building. Supporters of this governance model argue that transnational actors help build a common language and identify comparative indicators.

Not everyone is convinced about the positive impact of hybrid governance. Many experts have raised ethical questions about the legitimacy and transparency of a type of global governance that depends on private, nonelected, international organizations not directly accountable to voters or taxpayers. They contend that delegating policy design to consultants and nongovernment agencies reinforces neoliberal practices. Critics argue that agencies and consultants may pursue their own agendas without sufficiently adapting initiatives to the specific conditions of a given context. They claim that philanthropists, think tanks, and agencies are increasingly orchestrating climate action to protect private interests instead of responding to the real needs and expectations of the most vulnerable.

Orchestration here refers to a mode of indirect governance whereby an institution attempts to influence a target population through intermediaries using noncoercive means.[1] Orchestration often leads to "greenwashing" and adaptation initiatives that may comply with international sustainability or investment standards but also result in secondary effects such as social inequity. Others question the on-the-ground impact of orchestration and the ethical consequences of implementing foreign concepts. They argue that consulting services are often limited in time (their contracts typically end following the delivery of reports, guidelines, pathways, road maps, and checklists) and rarely include disciplined, long-term implementation solutions, monitoring, or follow-up. These limitations ultimately foster cities' ongoing dependence on external expertise. Finally, other critics contend that even when these experts make changes within municipalities, often by creating new climate or disaster-risk departments or units, these structures quickly become drained of expertise, resources, or administrative mechanisms required to implement long-term change.

LORENZO CHELLERI: INTERNATIONAL AGENCIES, CONSULTANTS, AND OTHER "ORCHESTRATORS" TRULY HELP CITIES REDUCE CLIMATE-RELATED RISKS

By reflecting on my career path, I would like to share how and why my response to the debate on the efficacy of orchestrators has changed over the years from "not at all" to "of course they help."

When I started my PhD, I was so skeptical about resilience that, along with some colleagues who shared similar concerns, I cofounded a research network that challenged the usefulness of the concept. We felt that orchestrators' grand claims about "building resilient cities" were facades to cover up large urban infrastructure initiatives. In addition, we called out their "business-as-usual" agendas for inducing and perpetuating "green or climate gentrification." Our first paper asserted that "resilience is not a normative positive concept," and several of my blog posts emphasized the fallacy of urban resilience.[2] Our research drew on several case studies of resilience, including evidence of desalination plants reducing drought risks while increasing energy consumption, which demonstrates how the rising water tariff pollutes the environment; massive solar power plants built in arid regions in need of water; how the costly reconstruction of a safer, flood-proof New Orleans displaced vulnerable populations; the contested green belt in Medellín; and many other examples of climate and green gentrification in the Global North.[3]

While writing these articles on resilience and risk trade-offs, I became more acquainted with the work done by many of the "orchestrators" I had previously criticized. I saw the valuable reports, useful tools, thoughtful recommendations, and critical thoughts they had developed over many hours of hard work, now abandoned on politicians' desks. I then recognized the complexities in implementing their recommendations. For example, many colleagues working with city councils experienced backlash for unsuccessfully lobbying politicians to consider climate change or sustainability within broader initiatives. The further I advanced in my career, the more I wondered who was responsible for not listening to the recommendations of thousands of scientists of the Intergovernmental Panel on Climate Change (IPCC). I slowly realized that many of these reports, guidelines, tools, and frameworks developed by "the orchestrators" were welcomed within policy statements but then rejected by policymakers during implementation.

At this point, I began to see that a growing critical mass of people was fighting for resilience and disaster risk reduction. We cannot blame the concept of sustainability, nor its advocates, for greenwashing. Similarly, resilience is not responsible for "safety-washing." Realizing how bad implementation, caused by stakeholder interests and behaviors, might thwart well-intentioned policies, I started to understand and value the committed work of consultants, associations, and city networks in advocating for best practices.

I invite you to reflect on the work of these so-called orchestrators. Consider how thousands of small cities have benefited from adaptation-designed plans thanks to the support and commitment of their mayors, who overcame the inefficiency and blindness of their provincial, regional, or national governments. Think about how many cities and practitioners have benefited from having their best practices and initiatives amplified by supporters, creating opportunities for co-learning via organizational events and peer-learning programs. Finally, I ask you to contemplate how many professionals, students, politicians, and activists are stepping forward to advocate for resilience and disaster risk reduction after the work of these orchestrators. In my research career, I have witnessed how and why orchestrators help those who ask for advice and are committed to reducing climate risks.

CRAIG JOHNSON: INTERNATIONAL AGENCIES, CONSULTANTS, AND OTHER "ORCHESTRATORS" ARE NOT TRULY HELPING CITIES REDUCE CLIMATE-RELATED RISKS

In a world of fragmented global governance, municipalities and transnational city networks are often portrayed as saviors. Their proximity to urban populations and responsibility to provide essential urban services make them particularly well suited for representing communities whose needs are either lost or ignored in nation-state politics.[4] However, the notion that international agencies, paid consultants, and other policy orchestrators will help reduce the climate-related risks of vulnerable urban populations implies that they have an interest in doing so.[5] Focusing primarily on climate consultants and city networks, I make the case that transnational efforts to reduce urban climate risk are biased to protect the assets and interests of relatively affluent cities and urban elites.

It is essential to recognize that city networks are not all created equal. Donor-driven models bring together cities and city networks through time-bound payments tied to programmatic goals; an example is the now defunct 100 Resilient Cities program, which the Rockefeller Foundation funded and subsequently scrapped in 2019.[6] In contrast to donor-driven models, there are decentralized networks like ICLEI and United Cities and Local Governments (UCLG), whose climate adaptation programming has a much larger and longer history of grassroots collaboration.[7]

The corporate network model represents a third—and dominant—form of city network governance, whose connections to global finance and investment make them far more visible and influential in international climate policy networks.[8] C40 is, arguably, one of the most ambitious examples of corporate city-network governance. Funded by American billionaire Michael Bloomberg, the network claims to represent the interests of "nearly 100 world-leading cities collaborating to deliver the urgent action needed right now to confront the climate crisis."[9] Its collaborations with the international consulting firm Arup have produced many policy recommendations and reports that are ostensibly designed to build urban resilience to climate change.[10]

I can raise three significant concerns regarding this model of urban climate governance. First, the problems that networks like Arup and C40 are trying to resolve are geared primarily toward relatively affluent cities, whose populations do not represent the greater number of residents from secondary and low-income cities who face wide-ranging vulnerabilities to climate change.[11] Second, the solutions these networks offer (or sell) demand costly investments, such as flood-modeling systems and large-scale rainwater retention tanks that exceed many municipalities' financial and political capacity.[12] Third, this model does not focus sufficiently on policies that address factors causing systemic vulnerability, including land rights, labor rights, affordable housing, domestic violence, and systemic racism.[13]

In practice, corporate city networks work primarily with mayors, city officials, and teams of paid consultants to develop plans, metrics, and other solutions for reducing risk and exposure to extreme climatic events. They do not generally work directly or collaboratively with vulnerable populations whose access to affordable housing, living wages, clean water, food security, and other essential urban services remains limited.

LORENZO CHELLERI'S REBUTTAL REMARKS

I agree with Johnson's overall criticism of the potential bias of donor-driven or corporate city networks, as well as how such governance models largely disregard considerations of systemic vulnerability, and the priorities they place on protecting assets over, for example, investments in adapting the houses of the urban poor. However, I see two gaps in his arguments, which can lead to problematic generalizations.

First, Johnson's points to lack of specificity. On the one hand, if big engineering firms like Arup and AECOM, among others, are looking to profit from climate action, or if the Rockefeller Foundation has started to build a group of "lighthouse cities" to launch its climate change program, we can expect outcomes based on short-term gains in technology and infrastructure to take precedence over addressing structural vulnerabilities. On the other hand, critiquing orchestrators' biases while focusing only on the C40 or 100 Resilient Cities program initiatives emphasizes a narrow perspective on the issue. Alongside orchestrators, other significant players like ICLEI work with small towns and cities to coordinate volunteer-based events, such as debates and tutoring sessions, that can disseminate knowledge and provide peer collaboration. I see substantial value in the work of such orchestrators, who raise awareness about climate risks and provide tools to help cities move toward resiliency beyond the solutions implemented by giant engineering firms.

The second gap in Johnson's argument, which concerns the central topic of this discussion, is how issues of structural vulnerability, justice, and sustainability relate to climate risks. I agree with Johnson's point that few investments prioritize vulnerable populations. In an unsustainable world divided by increasing inequality, such behaviors and outcomes are to be expected. But this debate does not ask, "Do orchestrators truly support cities in addressing social vulnerability to climate change?" or "Do orchestrators truly drive climate actions toward social justice?" If these were the questions posed, I would initially respond no. However, upon further reflection, I would say that even programs such as C40 and 100 Resilient Cities have promoted a broader conversation on climate resilience, equity, and justice.

PARTICIPANT'S COMMENTS

Steffen Lajoie, PhD candidate, Faculty of Environmental Design, Université de Montréal, Canada

I think that the opportunity to be an "orchestrator" is open to anyone who wants to take it. This debate has raised legitimate concerns about the democratic accountability of these actors. But in many urban contexts, political institutions are not accountable to urban inhabitants. This tension is felt most intensely by vulnerable populations experiencing poverty, marginalization, and insecurities. Where civic actors or activists are fighting for a voice in urban coproduction, international orchestrators can offer solidarity and provide legitimacy. The same is true for bureaucratic employees working to influence local policy from within institutional structures. Ambitious and imposed international policy that is socially just can provide additional wiggle room to positively influence policy and maybe even action.

The fact that we are having this debate on an international academic platform is indicative of the power leveraged by socially just approaches to disaster risk reduction, vulnerability reduction, and adaptation over the past fifty years. The opening argument provides a strong example of how professional activists have succeeded in integrating in and influencing the international process of orchestration. As a result, they have built power for themselves and for those most vulnerable to the climate crisis. Yes, the international orchestration phenomenon must be criticized and influenced by those fighting for social justice as community activists, volunteering professional allies, civic organizations, local or regional bureaucrats, consultants, and academics, among others.

In my view, it is preferable to have a shoddy orchestration platform in place that can be critiqued and coopted for building social justice than none. Former outsider activists can be employed to consult and influence global to local policy. Throughout history and in many contexts today, much of this work has been and remains radicalized if not criminalized, not to mention unremunerated.

There is a substantial amount of knowledge and capacity building in place compared to ten years ago, putting people at the center of climate initiatives. Arup's proposed City Resilience Framework has also been made freely available to the public since its launch. Dozens of cities not accepted to the 100 Resilient Cities program could still begin to build a resilience strategy using, or inspired by, the Arup website's reports, tools, and templates. After six years, their work on 100 Resilient Cities led to the formation of a new network of cities that does not rely on any donor-based mechanisms to implement resilience initiatives; Arup is now, through sharing tools and knowledge, widening its membership to allow any city to do resilience in its own way.

Orchestrators offer various supporting mechanisms to reduce risks, from investments to social impact–driven tools and service-driven instruments that improve the health of ecosystems in a limited number of accepted cities. However, knowledge-sharing solutions mean leaving behind short-term or donor-driven networks while empowering any city with the opportunities and resources to address structural vulnerabilities or the specific risks associated with resilience. Cities, rather than orchestrators, now bear the responsibility to address vulnerabilities.

CRAIG JOHNSON'S REBUTTAL REMARKS

I agree that many cities and city networks are doing excellent work to raise awareness, build capacity, share knowledge, and mobilize action on urban vulnerability, disaster risk reduction, and climate change. My comments on C40 are meant to highlight the new forms of corporate power currently shaping the contemporary landscape of urban climate governance.[14]

Strictly speaking, orchestration is a theory that aims to understand, at a global scale, the organization and projection of power. In the article I cowrote with Dave Gordon, we argue that city-network orchestration reflects both the proliferation of sub- and nonstate actors and the fragmentation of traditional multilateral governance institutions. Particularly important in this regard is "emergent orchestration," which describes the growing influence of industry standards, metrics, and performance indicators in ranking and comparing cities.

That cities are viewed positively by investors, businesses, employers, and bond-rating agencies matters. These actors manage, compare, and orchestrate perceptions of risk, urban policy narratives such as "livability," and capital flows that remain crucial for improving the health and prosperity of cities. In an age of acting locally and thinking globally on climate change, effective policies for reducing urban emissions and improving ambient air quality provide important metrics that connect cities to private and multilateral investment platforms like the Cities Climate Finance Alliance or the World Bank's City Creditworthiness Initiative.[15]

So, do transnational networks and management consultants make cities less vulnerable to climate risk? We must examine the relationship between management consulting and transnational city networks. Management consulting implies a wide range of professional services aimed at helping public, private, and non-profit institutions achieve their core goals and objectives. In the words of the American consulting firm Bain & Company, management consulting focuses on replacing visions and dreams with "facts and common sense."[16]

However, critics have stated that management consultants hired to develop urban climate governance plans prioritize large and valuable assets—housing, businesses, and infrastructure—over reducing chronic risks and vulnerabilities among marginalized urban populations.[17] This is partly limited by the challenge of collecting accurate and timely data, but it also reflects the limitations of making and managing disaster risk reduction priorities through management consulting approaches.[18]

Take, for instance, the Global Covenant of Mayors, a city network that is much larger and more diverse than C40.[19] However, like C40, it has worked closely with Arup to develop its Climate Risk and Adaptation Framework and Taxonomy (CRAFT), which "enables cities to perform robust and consistent reporting of local climate hazards and impacts."[20] In one sense, this taxonomy offers an important benchmark for city planners and officials to use in reducing urban climate risk. In another, it provides a metric that banks, businesses, and bond-rating agencies can use to assess the financial risk of insuring or investing in cities with similar hazard risk profiles.

Like any information technology, risk assessment platforms such as CRAFT create new systems, cultures, and capabilities whose applications are difficult to predict and control. However, adopting a management consulting approach to address risk reduction priorities has implications for democratic and civic engagement, which I will address in my concluding remarks.

PARTICIPANT'S COMMENTS

Maria Ikonomova, PhD candidate, Centre for Sustainable Development, Department of Engineering, University of Cambridge

Taking a stance that orchestration efforts should address structural vulnerabilities leads to the question of whether existing orchestration networks can achieve this task. Johnson's view is that the goals of existing orchestration networks are not aligned to focus on the mitigation of vulnerabilities. Lorenzo also shares his experience that policymakers often do not implement guidelines, tools, and frameworks developed by orchestrators; however, if this is the case, can orchestrators help cities target more challenging policies to address structural vulnerabilities?

The ways institutional orchestrators allocate funding and other resources such as human capital indicate their values. The wider climate change adaptation literature shows that climate adaptation funding models at an international level often do not lead to funding based on vulnerability and recipient needs, and that allocation decisions can also be made based on other political or economic factors.

While city orchestrators can provide human resources, they do not necessarily allocate funding for climate change adaptation initiatives; rather, they connect cities with funding institutions that distribute such aid. I think we must not only focus on the values of orchestrators working for cities but also examine the values and priorities of cooperating institutions such as development banks and donors. The values of these actors also influence the priorities framing climate change adaptation aid to cities.

LORENZO CHELLERI'S CLOSING REMARKS

The insights in Johnson's arguments highlight how it is necessary to fix a critical lens on the activities of major consulting firms. It is especially important to observe how these firms' reports and tools ground a metrics-based understanding of investments following only selected needs. I agree with Johnson regarding the potential danger of such corporate influences. In my recommendations for resilience, I highlighted aspects of orchestration that should be targeted, like investments and improvements, and those that should be neglected. We should keep this distinction in critical focus when debating the role of orchestrators in resilience issues, and how their wielding of power and influence is perceived. Perhaps their influence is less widespread than previously imagined; if so, we may better understand how much these stakeholders support one another.

Scientific literature provides dozens of assessments and evaluations of different networks' tools and programs: studies on the roles of ICLEI as a connector, mediator, translator, and educator; the impact of the Rockefeller 100 Resilient Cities program; and how justice considerations were integrated into this latter program.[21] This literature includes a three-year research project on climate resilience city networks, in which I had the pleasure to participate. We reviewed networks, associations, and initiatives, including groups of mayors, ICLEI, Climate Alliance, UNISDR, 100 Resilient Cities, Energy Cities, C40, and Eurocities, among others.[22] After conducting many interviews to understand how knowledge transfer and learning nourished these initiatives, we recently published a synthesis of our findings in a paper titled "City-to-City Learning Within Climate

City Networks: Definition, Significance, and Challenges from a Global Perspective."[23]

In this paper, we unveiled how, within these networks, "lighthouse" or "frontrunner" cities—those that often worked with the big firms cited by Johnson in his remarks—were intended to provide solutions and models to be followed by other cities. In practice, it is difficult for cities to follow these examples. As Johnson explains, large groups lobby to standardize the implementation of disaster risk reduction and climate resilience, usually working with major cities. Many ISO standards for sustainability now exist beyond CRAFT, including a Resilient Cities ISO.[24] However, we found that representatives from small cities, towns, and even "follower cities" generally admitted that the final aim of their involvement in orchestrators' activities is simply networking. By networking with other practitioners, they shared mistakes and tips on how to lobby for policy objectives—activities obscured from workshops, programs, tools, or frameworks that emphasize best practices. The 100 Resilient Cities framework, whose top-down approach represents the governance logic described by Johnson, also focuses on networking. In reality, cities faced with resilience issues will probably adapt and employ their own interests, projects, and local lobbies rather than the recommendations of major orchestrators.

On the one hand, despite top-down lobbyists pushing to drive policy implementation in cities, the cities themselves will ultimately decide how to proceed on climate and risk reduction. Conversely, even if Johnson is correct regarding the influence of lobbies, there are dozens of other networks, consultants, and organizations, such as Municipalities in Transition, the European Network for Community-led Initiatives on Climate Change and Sustainability (ECOLISE), and many others, that are also working closely with cities on sustainable, community-driven, justice-oriented climate resilience.[25] Therefore, orchestrators work according to various aims and perspectives that can genuinely help if cities wish to work with them. To address this debate, we must examine who drives risk reduction initiatives and in which direction.

CRAIG JOHNSON'S CLOSING REMARKS

Chelleri raises some excellent points about how decentralized networks like ICLEI and UCLG build capacity, disseminate knowledge, and foster collaboration among cities and municipalities. However, to suggest that the Rockefeller Foundation and C40 are a "small sample within a big, complex picture" understates the significant degree to which corporate-city networks are now shaping the policy agendas of many cities and other city networks. For example, ICLEI collaborates extensively with C40, particularly on its community-based protocol for measuring and reducing GHGs.[26] Although promoting collaborations among various governance networks and cities can increase synergies, ICLEI's long struggle to secure stable financing propels collaboration with corporate networks like C40.

This returns us to Chelleri's second point about risk and vulnerability. Although I agree that our debate focuses on how international organizations, consultants, and orchestrators reduce the risk of climate change, I disagree that risk and vulnerability can or should be separated so easily from such concerns. First, risk and vulnerability are not separate categories but interrelated processes whose impacts reflect the historical legacy of governmental and other institutional decisions, actions, and nondecisions.[27] Second, reducing risk often creates new vulnerabilities, especially when introducing infrastructures and technologies that dispossess people of their homes, livelihoods, and ways of living.[28] Third, the risk of losing one's home, livelihood, or life to disaster events is not evenly distributed among urban populations, reflecting the social and political processes that create and sustain structural patterns of vulnerability.

Critiques of risk and vulnerability bring us to the broader question of social justice. As introduced at the start of this chapter, corporate city networks can be criticized for promoting solutions that are not directly accountable to voters or taxpayers. Chelleri responds that city networks like C40 and 100 Resilient Cities encourage new debates about climate resilience, equity, and justice. However, the plans promoted and sometimes adopted by city leaders and municipalities are often biased in favor of protecting the wealthy and powerful. Indeed, a recent study of forty-three cities by Patricia Romero-Lankao and Daniel Gnatz found that C40 adaptation plans tend to prioritize "techno-infrastructural and economic investments" at the expense of food insecurity, energy poverty, and water scarcity.[29]

The recognition that city network governance is contingent upon the power and wealth of corporations, consultants, and their benefactors draws our attention to the productive forces driving the political economy of labor, capital, and accumulation at the urban scale. In the context of climate change, this raises questions about how cities can act effectively or independently without private orchestrators such as Bloomberg, C40, and Arup. In contrast to the participatory models adopted by decentralized city networks, corporate city networks and consultants work in a policy vacuum that understates or ignores the forces driving urban vulnerability and equality, including rising land values, inadequate housing, and precarious work. Corporate city climate solutions offer plans for measuring and improving urban services without addressing the systemic factors perpetuating vulnerability to climate change.

PARTICIPANT'S COMMENTS

Lisa Hasan, PhD candidate, Faculty of Environmental Design, Université de Montréal, Canada

In response to the question asking whether the presence of orchestrators is morally neutral, my short answer is no.

A high degree of transparency is needed to appreciate the ethical position of any organization. As a form of indirect governance whereby an orchestrator steers a target through an intermediary, orchestration inherently poses challenges to transparency. While all orchestrators call for greater transparency from their members—by the publication of plans or the disclosure of data—they do not always apply the same standards of transparency to their own governance. Thus, establishing where a city network falls on the ethical spectrum is a herculean task for any city. The only cities for which this determination is facilitated are those that also act as orchestrators. Cities like New York and Toronto have benefited from their mayors' occupying key positions within C40 and the Global Covenant of Mayors. Orchestrator cities help define a "shared language" and are given more leeway in how they interpret its application. So, perhaps orchestrators are not inherently fair or unfair, but rather fairer for some cities and less fair for others.

Cities that rely on funds from orchestrators to implement risk reduction measures may not be able to question problems or proposed solutions. Waiting to see whether these organizations act ethically before engaging may not be a viable option. Perhaps one way for cities to test the ethical orientation of orchestrators is to see to what extent they are open to allowing cities and communities to challenge their definitions and proposals before agreeing to engage. All international organizations suffer from inefficient behaviors, but are the gaps between stated values and actions intentional or inadvertent?

While Chelleri's optimism results from the recent increase in public debates on climate resilience, equity, and justice, I question the growth of discourse as a measure of progress. Determining whether inefficient behaviors are going unnoticed, are being purposefully maintained, or are simply hard to resolve is no simple task. It requires finding evidence of individuals trying to overcome these obstacles or actively maintaining them. I have found it useful to ask: Who stands to gain or lose from ineffective behaviors? Does evidence exist regarding efforts to overcome the gaps between values and impacts? Have these efforts been met with resistance? Who is resisting them, and why?

THE MODERATOR'S CLOSING REMARKS: THE EFFICACY AND MORAL VALUE OF "ORCHESTRATING" DISASTER RISK REDUCTION IN TIMES OF GLOBAL WARMING

International agencies, consultants, and other orchestrators increasingly influence climate action, disaster risk reduction, and urban policy. But their influence is constrained by both inefficiencies and fragile ethics. Are orchestrators part of the climate problem or solution?

This debate received divided responses among participants. A slight majority—about 52 percent—believe that orchestrators of urban policy are helping cities reduce climate-related risks (figure 8.1). During the debate, they found several advantages in hybrid governance and networking. Participants also highlighted several drawbacks to, and unresolved issues in, the orchestration of urban policy, including inefficacy in implementation and limited attention to ethical considerations.

At a time when we need to "think globally and act locally," our participants see hybrid forms of governance as crucial ways to deal with global problems while still responding to local citizens and social groups. For Lorenzo Chelleri and many others, international agencies, consultants, and other orchestrators provide tools and spaces to enact positive change. They allow municipal governments to network, share experiences, learn, and disseminate results. By creating tools and spaces for debate and engagement, international stakeholders provide alternatives to national programs and to the frequent inaction of politicians and agencies within central governments. Orchestrators connect ideas, metrics, and tools in ways that can have an impact on local policy. In addition, transnational actors help build a common language and identify comparative indicators. Cities can implement change only when decision-makers understand, measure, and compare urban conditions.

It is important to note that even defenders of hybrid forms of governance find it difficult to implement change, convince elected officers, and mobilize resources toward risk reduction for the most vulnerable. Reducing structural vulnerabilities is a challenge even for powerful alliances and city networks.

Orchestrators often cannot, or prefer not to, influence entire governance processes, from diagnosis and policy drafting to implementation, monitoring, and evaluation.

For Craig Johnson and other critics of orchestration, the problem lies not necessarily in the inefficiency of orchestrators' actions but in their moral value. To critics, the tools and spaces orchestrators adopt for change are not morally impartial. Instead, they find that hybrid forms of governance suffer from a series of pathologies. While they agree that cities, especially small ones, can benefit from the external aid and innovations required to tackle the challenges of climate change, they do not believe that the contributions offered by international agencies and consultants are risk-free. The instruments created and promoted by orchestration—including metrics, certifications, and evaluation techniques, among others—implement a certain vision of resilience, sustainability, adaptation, risk reduction, and progress. They are also based on notions from other global contexts, which are not always adapted to local conditions. As orchestrators grow in power and influence, municipalities are less able to challenge their narratives, concepts, language, and tools. The process of orchestration thus delegitimizes local ways of approaching difficult issues.

Risk also arises when cities rely on orchestrators to develop short-term climate solutions rather than address structural vulnerabilities through more complex policy measures. While defenders of orchestrators praise their development of frameworks, guidelines, pathways, and other policy instruments, critics are wary of their ability to manipulate actions, concentrate power, and control climate and risk reduction discourse.

For critics, conflicts of interest accompany many resilience strategies. The same companies that "sell" resilience plans and consulting services also fuel research and develop tools and metrics for resilience. Are international orchestrators contributing to local capacity building or simply creating new forms of economic and political dependence? Johnson concludes with "the recognition that city network governance is contingent upon the power and wealth of corporations and consultants; their benefactors draw our attention to the productive forces that drive the political economy of labor, capital and accumulation at the urban scale."

Ultimately, the debate over the role of orchestrators raises a key question: Why do cities lack the internal resources and skills required for disaster risk reduction? In many cities, neoliberal policies based on austerity measures and downsizing have dismantled or reduced internal services. Cities are, therefore, increasingly forced to contract with external experts to manage disaster risk reduction. Because of their insufficient power and agency, some cities have come together to voice their concerns globally, seeking a space for action and change beyond the control of states and provinces.

We are uncertain whether the results of this debate mean that participants believe that orchestrators' agency can or should be improved—this is perhaps a question for a future debate. But we know that given the diverse forms of orchestrator networks, some network models may be better suited than others to help cities build their adaptive capacities and reduce structural vulnerabilities. Perhaps some orchestrator networks and tools should be redesigned to help cities address such concerns. Finally, given comments and votes, most participants in this debate see structural vulnerabilities as resolvable only through the collaboration of citizens, civil society, cities, governments, and international stakeholders.

This debate has produced a fascinating analysis of the benefits and drawbacks of current forms of city governance. Panelists

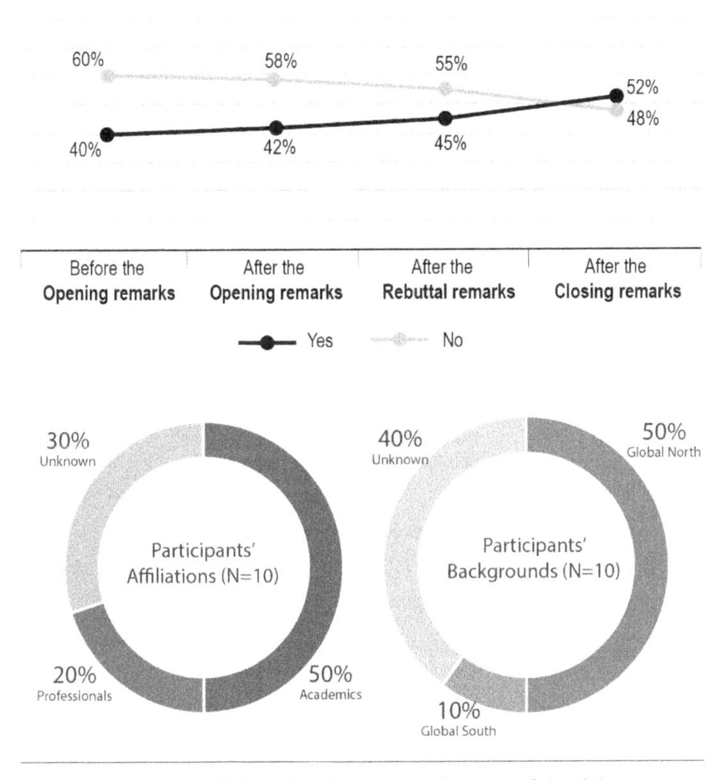

FIGURE 8.1 (*Top*) Results of votes at each stage of the debate. (*Bottom*) Distribution of affiliations and backgrounds among participants in the discussion.

and participants have guided us to more than twenty pertinent references—scientific articles, reports, and websites—that can inform this important debate. Additionally, public opinion on the role of orchestrators changed over the course of two weeks. At first, 60 percent of participants believed that the actions of orchestrators do not help cities reduce climate-related risks. After a few days, 52 percent of participants agreed that orchestrators did help cities.

NOTES

1. Kenneth W. Abbott et al., eds., *International Organizations as Orchestrators* (Cambridge: Cambridge University Press, 2015); David J. Gordon and Craig A. Johnson, "The Orchestration of Global Urban Climate Governance: Conducting Power in the Post-Paris Climate Regime," *Environmental Politics* 26, no. 4 (2017): 694–714.
2. Lorenzo Chelleri and Marta Olazabal, *Multidisciplinary Perspectives on Urban Resilience* (Bilbao, Spain: Basque Centre for Climate Change, 2012); Lorenzo Chelleri, "The Urban Resilience Fallacy: Gaps Between Theory and Practice," UGEC Viewpoints: A Blog on Urbanization and Global Environmental Change, October 11, 2016, https://ugecviewpoints.wordpress.com/2016/10/11/the-urban-resilience-fallacy-gaps-between-theory-and-practice/.
3. Isabelle Anguelovski et al., "Why Green 'Climate Gentrification' Threatens Poor and Vulnerable Populations," *PNAS* 116, no. 52 (2019).
4. Benjamin R. Barber, *Cool Cities: Urban Sovereignty and the Fix for Global Warming* (Hew Haven, CT: Yale University Press, 2017).
5. David J. Gordon and Craig A. Johnson, "The Orchestration of Global Urban Climate Governance: Conducting Power in the Post-Paris Climate Regime," *Environmental Politics* 26, no. 4 (2017): 694–714.
6. The Rockefeller Foundation pioneered 100 Resilient Cities to help more cities build resilience to the physical, social, and economic challenges; see https://www.rockefellerfoundation.org/100-resilient-cities/.
7. United Cities and Local Governments (UCLG) is a global network of cities and local, regional, and metropolitan governments; see https://www.uclg.org/.
8. C40 Knowledge, "Cop26: What the Glasgow Climate Pact Means for Cities," 2021, https://www.c40knowledgehub.org/s/article/COP26-What-the-Glasgow-Climate-Pact-means-for-cities?language=en_US.
9. C40 Cities, https://www.c40.org/.
10. C40 and Arup, "Deadline 2020: How Cities Will Get the Job Done," 2016, https://c40.my.salesforce.com/sfc/p/#36000001Enhz/a/1Q000 0001pEu/DMsADbdqrK6WCGBFhigJZ_MoEDgUboAFG7W gHuJTnNw; Desiree Bernhard et al., "Focused Acceleration: A Strategic Approach to Climate Action in Cities to 2030," McKinsey

Center for Business and Environment, November 2017, https://www.mckinsey.com/~/media/mckinsey/business%20functions/sustainability/our%20insights/a%20strategic%20approach%20to%20climate%20action%20in%20cities%20focused%20acceleration/focused-acceleration.pdf.

11. Secondary cities are defined as small or midsize cities, https://www.citiesalliance.org/themes/secondary-cities.
12. C40 Knowledge, "How to Reduce Flood Risk in Your City," November 2021, https://www.c40knowledgehub.org/s/article/How-to-reduce-flood-risk-in-your-city?language=en_US.
13. Patricia Romero-Lankao and Daniel Gnatz, "Risk Inequality and the Food-Energy-Water (FEW) Nexus: A Study of 43 City Adaptation Plans," *Frontiers in Sociology* 4 (2019): 31.
14. Craig A. Johnson, *The Power of Cities in Global Climate Politics: Saviours, Supplicants or Agents of Change?* (London: Palgrave Macmillan, 2018).
15. The Cities Climate Finance Leadership Alliance works for urban climate action to mobilize the resources needed for a low-carbon transition in cities, https://citiesclimatefinance.org/; World Bank Group, "City Creditworthiness Initiative: A Partnership to Deliver Municipal Finance," 2022, www.worldbank.org/en/topic/urbandevelopment/brief/city-creditworthiness-initiative.
16. See Bain & Company, "What Is Management Consulting?," https://www.bain.com/careers/what-is-management-consulting/.
17. Kathryn Davidson, Lars Coenen, and Brendan Gleeson, "A Decade of C40: Research Insights and Agendas for City Networks," *Global Policy* 10, no. 4 (2019): 697–708.
18. David Satterthwaite and Sheridan Bartlett, "The Full Spectrum of Risk in Urban Centres: Changing Perceptions, Changing Priorities," *Environment and Urbanization* 29, no. 1 (2017): 3–14; Ilan Kelman, *Disaster by Choice: How Our Actions Turn Natural Hazards Into Catastrophes* (Oxford: Oxford University Press, 2020).
19. See Global Covenant of Mayors for Climate & Energy, https://www.globalcovenantofmayors.org/.
20. Climate Risk and Adaptation Framework and Taxonomy (CRAFT), https://www.globalcovenantofmayors.org/resource/climate-risk-and-adaptation-framework-and-taxonomy-craft/.

21. Niki Frantzeskaki et al., "The Multiple Roles of ICLEI: Intermediating to Innovate Urban Biodiversity Governance," *Ecological Economics* 164 (2019): 106350; Sahar Zavareh Hofmann, "100 Resilient Cities Program and the Role of the Sendai Framework and Disaster Risk Reduction for Resilient Cities," *Progress in Disaster Science* 11 (2021): 100189; Joanne Fitzgibbons and Carrie L. Mitchell, "Just Urban Futures? Exploring Equity in '100 Resilient Cities,' " *World Development* 122 (2019): 648–659.
22. Eurocities is a network of more than two hundred cities in thirty-eight countries, https://eurocities.eu/.
23. Wolfgang Haupt et al., "City-to-City Learning Within Climate City Networks: Definition, Significance, and Challenges from a Global Perspective," *International Journal of Urban Sustainable Development* 12, no. 2 (2020): 143–159.
24. ISO, *ISO 37120: Sustainable Cities and Communities—Indicators for City Services and Quality of Life* (Geneva: International Organization for Standardization, 2018; ISO, *ISO 37123: Sustainable Cities and Communities—Indicators for Resilient Cities* (Geneva: International Organization for Standardization, 2019).
25. Municipalities in Transition (MiT) develops and tests a structured way for municipalities and local governments to connect with their communities and respond to the great challenges, https://municipalities intransition.org/; ECOLISE, https://www.ecolise.eu/.
26. Wee Kean Fong et al., "Global Protocol for Community-Scale Greenhouse Gas Inventories: An Accounting and Reporting Standard for Cities," n.d., https://ghgprotocol.org/sites/default/files/standards /GPC_Full_MASTER_RW_v7.pdf.
27. Ilan Kelman, "Lost for Words Amongst Disaster Risk Science Vocabulary?," *International Journal of Disaster Risk Science* 9, no. 3 (2018): 281–291.
28. Benjamin K Sovacool, Björn-Ola Linnér, and Michael E Goodsite, "The Political Economy of Climate Adaptation," *Nature Climate Change* 5, no. 7 (2015): 616–618.
29. Romero-Lankao and Gnatz, "Risk Inequality and the Food-Energy-Water (FEW) Nexus."

9

ON REGULATIONS

Should Governments Devise and Enforce Standards for Low-Cost Housing in Developing Countries?

WITH EDMUNDO WERNA AND BRIAN ALDRICH

In 2006, the Colombian housing minister announced that his government "would never subsidize a single housing unit having less than two bedrooms." This comment summarizes what many academics, decision-makers, and politicians believe: that minimum standards for low-cost housing in developing countries are necessary to guarantee everyone's right to an adequate standard of living. They note that millions around the world live in life- or health-threatening conditions, substandard housing, and neighborhoods associated with housing stock dilapidation, overcrowding, poor infrastructure, lack of ventilation, and other conditions that do not uphold their human rights and dignity.[1] They often claim that most informal settlements are unfit for human living and that governments and authorities must adopt building standards that guarantee basic infrastructure, appropriate location, proper materials and design, and sound structural and construction techniques.[2] The World Bank and the Global Facility for Disaster Reduction and Recovery (GFDRR), for instance, launched in 2015 the Building Regulation for Resilience Program (a report, feature story, and a video are available).[3] Researchers and policymakers argue it is necessary to define and enforce minimum standards across scales of building—local, national, and international—to prevent unscrupulous or illegal construction, remedy poor quality housing, achieve safety and sustainability, reduce inequality and discrimination, and secure housing tenure.[4]

Opponents of minimum standards for low-cost housing question their usefulness and effectiveness. They claim that low-cost housing is characterized by progressive or incremental development, as limited resources force families to build their housing and communities in stages. Thus, homeowners act according to their needs, enabling them to synchronize their investment in housing in rhythm with social and economic changes.[5] Opponents also argue that adopting minimum standards for low-cost housing exacerbates housing crises by disregarding the economic and social needs of poor urban dwellers, for whom formal construction often remains unaffordable. Some analysts contend that minimum standards often overlook environmental conditions, local identity, culture, and traditions, thereby causing irreversible physical and social impacts on communities. Finally, critics contend that enforcing minimum standards for low-cost housing encourages more informal construction and, in many cases, unnecessarily forces the poor into illegal living conditions, which in turn exacerbates vulnerability.

EDMUNDO WERNA: GOVERNMENTS SHOULD DEVISE AND ENFORCE STANDARDS FOR LOW-COST HOUSING

A plethora of arguments stand against regulation, from stating that the poor should be allowed to live in substandard housing because they cannot afford better alternatives, to supporting the general deregulation of housing markets, including the non-use of standards. While some of these arguments hold, especially those that protect the interests of the poor, we should be wary of dismissing standards altogether. Standards for housing exist for a reason; some examples are outlined below.

First, architectural standards, such as the minimum level of ventilation, the volume of air in each room, lighting, smoke exhaustion, and sanitary devices, are regulated to avoid potential health risks to residents. Cases of housing-related accidents, diseases, and deaths unfortunately abound. For similar reasons, regulations prevent the use of building materials such as asbestos, toxic paints, and other harmful products. Second, construction standards control the quality of building processes and structural resistance. Cases of individual houses or apartment buildings collapsing during or after construction remain frequent. Third, urban design and labor standards vitally interface with housing standards, although they may be beyond the scope of this debate.

Problems will continue if standards are not respected. It is unreasonable to accept a public agency opting for production (whether subcontractor or government built) of substandard housing units at lower cost than standard ones if this choice harms residents' health.

Producing below-standard housing does not necessarily result in a better price for consumers. The savings gained by cutting building costs may be used to increase the profits of construction enterprises and real estate investors. The private sector must compete within a framework of healthy regulations; otherwise, housing standards and working conditions may negatively amplify each other, producing an ever-lowering standard of housing and ever-worsening working conditions, in a race to the bottom.

Although these arguments all favor housing standards, it is important to acknowledge that building regulations should be applied with caution because many residents cannot afford higher-quality housing. Housing quality should not correlate to purchasing power. Regulations have frequently been used as excuses to bulldoze substandard housing, leading to evictions and prioritizing real estate development interests over disadvantaged residents. Standards and regulations should be used in favor of, rather than against, residents' health and interests. Governments should act as advisers for those who cannot afford better housing by promoting the benefits of housing standards, finding ways to upgrade existing developments, or redistributing taxes to make standard housing more affordable.

BRIAN ALDRICH: THE CHALLENGES OF ENFORCING GOVERNMENT STANDARDS FOR LOW-COST HOUSING

The demand for adequate housing is increasing in low- and moderate-income countries.[6] The result is overurbanization: cities are no longer able to provide their growing populations with adequate jobs or services such as housing. Percentages of residents living in illegal housing or squatting on land they do not own can vary from 0 percent (Singapore) to 60 or 70 percent (Lagos).

Overpopulation is primarily caused by low levels of economic development, as governments cannot pay for low-income housing with low gross financial performance. International agencies imposing strict economic standards, participation in the global economy, international markets for domestic products—any or all these factors can limit economic growth by restricting government spending and limiting growth. Economic ups and downs have the most significant impact on the poor and the emerging middle class.

Several conditions limit access to standardized housing, among them the availability and price of land, legal title, and conflicting systems of property ownership. Also important is the power of economic and political elites to manipulate regulations governing access to land. The resulting impediments may take many forms, including colonizing unused or vacant land, excessive bureaucratic requirements, kleptocracy or corruption in state bureaucracies, poorly designed or inadequate enforcement of standards, autocratic or single-party control of access to housing, and extensive patronage systems. Top-down planning by the state does not generally benefit the poor who face one or more of these limiting conditions when seeking housing.

If the poor must live in cities under these conditions, they must accept whatever housing is available. They are often left to fend for themselves by building or renting marginal housing wherever space is available. Typically, these sites are on floodplains, hillsides, or other marginal areas, and are characterized by crowding, overbuilding, and a lack of reliable electricity, water, and sewage services. The poor may live under bridges, on city dumps, on toxic land near factories, on desert lands surrounding a city, on vacant government or private land—anywhere they can informally live or work. Underprivileged residents cannot standardize their housing conditions when residing in illegal settlements.

The poor generally occupy low-paying jobs tied to specific city sectors, which makes commuting from outlying relocation sites prohibitively expensive and time-consuming. They are typically cut off from other urban social services as well.

EDMUNDO WERNA'S REBUTTAL REMARKS

On September 27, 2013, a residential building collapsed in Mumbai, killing sixty-one dwellers and injuring thirty-two others.[7] According to investigation analyses, the disaster occurred because of the substandard and illegal construction of an extra floor. This is far from the only negative impact of neglecting housing standards.

The arguments presented by Aldrich provide a solid description of the plight of the poor unable to afford appropriate housing. He correctly states that access to regularized housing remains limited and inequitable for many reasons. However, this does not mean that housing standards should be ignored altogether. The disaster in Mumbai reminds us that concrete actions must be taken to develop building regulations to prevent future problems.

According to the UN-Habitat and WHO report on Urban Health, "even if the housing structure is durable, lack of adequate ventilation, air-conditioning, heating and use of hazardous building material can cause acute effects on health and comfort." It also stated that "inadequate urban housing blights the health of billions of people worldwide. . . . Many slum dwellers live in houses with . . . poor-quality roofs and walls constructed out of waste materials. . . . These houses do not provide proper protection against inclement weather, parasitic infections, or unwelcome human intruders."[8] WHO estimates that housing accounts for more than a hundred thousand deaths annually, even in developed countries, resulting from hazards such as noise, dampness, indoor air quality, cold, and home safety. These problems are technically preventable. Building standards provide necessary benchmarks for preventive action. If acceptable limits and regulations are not in place, how can appropriate housing be designed, built, and promoted?

PARTICIPANT'S COMMENTS

Kirti Joshi, Professor, Lumbini Technological University, Nepal

It is not a question of whether regulations are needed, but what type of regulations are required. How does one ensure that desired results are achieved, and regulations best implemented?

Following the 2015 earthquakes in Nepal (which caused nine thousand casualties and the collapse of six hundred thousand buildings), the government published a set of building plans to be followed by earthquake-affected households wishing to qualify for housing subsidies. But it took more than a year to prepare and approve such plans, and many of the solutions were unaffordable, given the meager budget of subsidies. Despite the government's good intentions to ensure risk-resilient housing, the plans disappointed affected households who had been waiting for housing subsidies for more than a year.

Although this example implies that some regulations can be inefficient and a waste of resources (including time), it does not discredit regulations as a whole. Building regulations are often imposed to account for negative externalities arising from the construction of buildings, such as traffic congestion, blocked sunlight or view, threat to neighboring buildings during seismic shocks, and other vulnerabilities. However, apart from the costs of implementation, building regulations can also both increase housing prices and decrease housing supply. In some cases, they can also encourage informal housing. There are certainly trade-offs involved in terms of the costs and benefits of regulations.

Although stringent regulations generally reduce welfare, some are required to protect occupants and their neighbors. They should be developed with consultation from affected communities, rather than imposed top-down.

The UN-Habitat and WHO publication acknowledges that global housing standards were initially based on health principles, yet these have been neglected, leading the authors to call for updated health standards in housing. We might also consider other standards, such as environmental ones, alongside their call. The built environment, which includes housing, is responsible for the most significant impact on the natural environment, affecting the health of residents and the entire planet. The risks of not paying attention to standards are wide-ranging.

As Aldrich points out, "when living in illegal settlements, underprivileged residents have no chance of standardizing their housing conditions." Although true, this statement should not stand as an argument to dismiss the importance of maintaining standards and regulations in housing. If the poor cannot afford standard housing, we should try to reverse this situation instead of turning a blind eye to their living conditions. To do this, we must combat poverty, as echoed in Aldrich's statements.

In my introductory remarks, I stated that the poor should not be marginalized or penalized for their purchasing power. Standards should not be used to excuse evictions or other actions against the poor. My remarks also noted how a lack of standards perpetuates increasingly debilitating risks. We may then need to rephrase the initial question of this debate to ask how to use housing standards for universal benefit instead of whether they are required. Should we forget about standards and hope that the disaster in Mumbai, as in many other places, does not happen again? Or should we take action against the causes leading to such incidents?

BRIAN ALDRICH'S REBUTTAL REMARKS

When I agreed to take the "con" position on the issue of the state imposition of housing regulations, I considered the implications of top-down regulations on low-income housing. The proponent of the "pro" position has softened the original proposition of the debate to acknowledge several alternative approaches to the main proposal. The top-down imposition of housing regulations on the poor under the conditions I have outlined has been a failure in most cases. Generally, it involves relocating residents, destroying existing housing, and rebuilding for few or no former residents of the area. Land is then used to build profitable, upper-middle-class housing or commercial centers. Usually, conglomerates in league with politicians control these new developments. Such operations marginalize poor communities in relocation settlements at the edge of growing cities, far from worksites, transportation, and social and educational services.

The presence of almost a billion poor people living in illegal, substandard housing is the result of a developing global society and the subsequent movement of the masses to large metropolitan areas. As the status of low-income groups around the world varies substantially, I suggest the following incremental approach:

First, adequate financial resources are needed. If metro areas do not generate sufficient wealth, then global organizations must provide funds and prioritize housing as a social necessity and human right.

Second, some form of organization of demand-based specialists is key to avoiding top-down housing. Teams of architects, planners, builders, and residents must collaborate to solve specific housing problems for low-income groups in defined areas. These teams must likewise have the power to negotiate housing regulations and develop enforcement procedures.

PARTICIPANT'S COMMENTS

John Plodinec, Distinguished Senior Fellow, Northeastern University's Global Resilience Institute, Boston

Everyone seems to agree that the principle of housing standards is correct; unfortunately, we live in an unprincipled world. The positive case studies cited in this debate can be compared to existence theorems in mathematics—they demonstrate that solutions exist. Sorting out the messy conditions under which these solutions work—and, more importantly, under what conditions "solutions" do more harm than good—is more difficult. In developed and developing countries alike, the working poor are effectively forced into less than desirable housing conditions whether because of corruption or developers chasing the almighty dollar.

I hesitantly voted "for" standards, in recognition that standards are ultimately best for the working poor. But I did so hesitantly in recognition that standards imposed from the top down, especially if implemented without consideration of local conditions or unintended consequences, can do more harm than good.

A bottom-up, site-specific, participatory approach helmed by a design team has a greater chance of providing housing services to low-income populations than a top-down approach. The Basic

Services for the Urban Poor (BSUP) model is being developed into a more effective general housing provider in India. Thailand similarly has the state-supported program Baan Mankong to negotiate small and large occupancy sites for groups of squatters.[9] Staffing this program has been successful, and resident participation has noticeably aided the rehousing effort. Brazil has developed the Instituto De Pesquisa Econômica Aplicada (IPEA), an outstandingly successful program for socially oriented housing, attracting high and varied participation levels; it is not only oriented toward solving housing issues but also aims to integrate residents into city life. The success of the project hinges on its flexible integration into local municipal policies, including those for the development of new and more effective forms of community organization.

EDMUNDO WERNA'S CLOSING REMARKS

"As safe as houses," so the traditional saying goes—but is it so? This saying refers to the risks of investing in real estate, but acquiring a house that causes harm to its residents can hardly be a safe investment, at least in terms of health and likely in terms of finances.[10]

In January 2016, a one-family detached house collapsed in the low-income neighborhood of Iraja, Rio de Janeiro, killing two residents and injuring three. The building was an incremental housing unit; its additions were unregulated, causing the disaster. Apart from the loss of human life, the investment was not a financially safe choice. The burgeoning literature on urban livelihoods rightly promotes housing as a critical asset for the poor, but this incident in Iraja is not a good example.

The multifamily apartment block mentioned in my rebuttal was also built incrementally without respecting standards, leading to similar disastrous results. I have presented additional data beyond the two examples from Mumbai and Rio, but Aldrich claims that I have softened my position on the debate while he has fully responded to the implications of top-down state regulations on housing. I disagree.

More analysis is needed to consider the complete picture of state regulations on substandard housing, which should include analysis of the benefits of nonregulation, the consequences of free markets, and the environmental, financial, health, and safety impacts on the poor.

My position in this debate never changed. I made it clear from the onset that I would defend the rationale of housing standards while emphasizing that the poor should not be penalized. To defend the rationale for implementing housing standards does not mean that one must protect an entirely top-down approach.

Aldrich's suggestions follow proposals by John Turner, Alan Gilbert, and many others.[11] Previously, I have argued similarly and even had the opportunity to practice an incremental approach in poor communities. However, I disagree with housing conditions that negatively affect the well-being of the residents. I ask again, what happens when standards are ignored? Aldrich does not address this question, but I hope the evidence I have provided makes a case for its consideration.

When Turner visited slum areas in Lima in the early 1960s, he was told they were "a problem." He replied that he did not see a problem, but a solution; he perceived such settlements as a means of empowering the poor. I also see a solution in such contexts, but not without problems, such as the risks of substandard construction. Aldrich has successfully argued for the development of small-scale housing. Still, I do not foresee their replicability on a large enough scale to solve the problem of adequate housing—our world needs millions of safe residences.

I condemn the financial inequity of the rich being able to afford housing well above necessary levels of comfort while the poor build their homes incrementally in risky, if not outright fatal, conditions. I reiterate the concrete proposals made in my introductory remarks about redistributing wealth through taxes. Public housing programs in countries mentioned by Aldrich provide safe housing for many people, such as Minha Casa Minha Vida in Brazil.[12] We may need to rephrase the original question of this debate to ask how standards can benefit everyone, rather than if they are required at all.

BRIAN ALDRICH'S CLOSING REMARKS

Werna and I are not really in disagreement in this debate. When I read his opening statement, it was clear that he did not support the proposition as stated but rather that the government should devise and enforce minimum housing standards. He made a persuasive argument that minimum building standards must be imposed to preserve the lives and health of residents. I agree with this position, but I wish to make clear that the full imposition of housing standards can be dangerous, impractical, and impossible.

Exceptions often prove the rule, some of which I describe in my recently published research on housing squatters and the urban poor in Southeast Asia's six megacities, including the city-state of Singapore and the former British colony of Hong Kong.[13] They are outstanding examples of how urban governments can, over fifty years or so, rehouse two to four million inhabitants by enforcing a unified program of housing regulations. Under close government regulation—no gum thrown on sidewalks, no planters placed on apartment ledges, and so on—Singapore has built new housing for 85 percent of the population. Almost the entire island has been transformed into a single functioning, structural, and renewable unit. Notably, the city-state has been governed by the same absolutist political party since its inception in the 1960s.

Hong Kong, which remained a British colony until the end of the last century, also wholly controlled the state provision of housing for the unhoused millions who moved there after years of turmoil in mainland China. Through a scheme in which the government sold sectioned land on the open market, three and a half million people were rehoused while any previous squatters were relocated. Since the colony's return to China, the state no longer manages access to these public lands, so it is difficult to remove those who surpass the income level required to reside in public housing.

Other Southeast Asian megacities such as Manila, Bangkok, Kuala Lumpur, and Jakarta have not successfully established housing standards for either health or safety. These standards are the minimum required and must be implemented incrementally to address the specific needs of low-income and squatter populations.

PARTICIPANT'S COMMENTS

Georgia Cardosi, Assistant Professor, Faculty of Environmental Design, Université de Montréal, Canada

Mechanisms of deregulation applied through the Structural Adjustment Programs in the 1980s caused state withdrawal in developing countries. Scholars agree that deregulation has negatively affected production in these countries and, consequently, the provision of infrastructure, services, and housing by the state. The lack of such basic services has led to high unemployment and lower incomes, the major causes of poverty, inequality, and marginalization, while encouraging informal arrangements in all sectors: economy, services, housing, and others.

Thus, housing is a process and not an object, as Turner argued, that should be developed concurrently with measures aimed at reducing poverty and promoting socioeconomic inclusion. Producing wealth should not be a focus, as access to wealth may be impeded by corruption, kleptocracy, and issues related to economic and political power highlighted by Aldrich. Rather, employment and income opportunities must be enhanced in tandem with increased access to affordable land and adequate housing.

Although regulations can produce drastic effects if not developed in consideration of the actual needs of the poor, the development of adequate housing standards for low-income households cannot be a choice—this is necessary to enhance social advancements. Rather, we should look at the debate in a fresh way; to quote Pamoja Trust, "realising housing for the poor requires concerted organizing. We must realize that we have to build people before we build houses."[a] This urgency applies to all stakeholders involved in the development of housing standards.

[a] Pamoja Trust, a nonprofit organization founded in 1999, is dedicated to promoting access to land, shelter, and basic services for the vulnerable, https://pamojatrust.org/.

THE MODERATOR'S CLOSING REMARKS: SOCIAL JUSTICE AND HOUSING REGULATIONS IN DEVELOPING COUNTRIES

This debate resulted in a resounding Yes in favor of housing standards in developing countries, provided they do not penalize the poor. Participants addressed the role of the government in regulating housing conditions in developing countries and the links between standards and social justice. They exposed the advantages and secondary effects of housing standards and raised questions about their effectiveness and applicability. In general, however, participants consistently claimed that governments should devise and enforce standards—about 60 percent of votes defended this position over the ten days of the debate. This stance is particularly relevant now that the World Bank—a traditional defender of deregulation in developing countries—and the Global Facility for Disaster Reduction and Recovery have released the ambitious "Building Regulation for Resilience" program.[14]

Generally, participants in this debate feared that, in the absence of standards, markets might experience a "race to the bottom," increasing vulnerabilities and penalizing the poor and urban populations in general. These results effectively answered the main question proposed by the debate, but also raised the following specific themes that require further analysis and consideration:

- *Scope*. For many specialists, the primary and perhaps only goal of housing standards should be to protect the health and safety of residents. In this view, governments should aim to enforce the minimum standards required and avoid regulations that may affect culturally and socially based modes of living, even

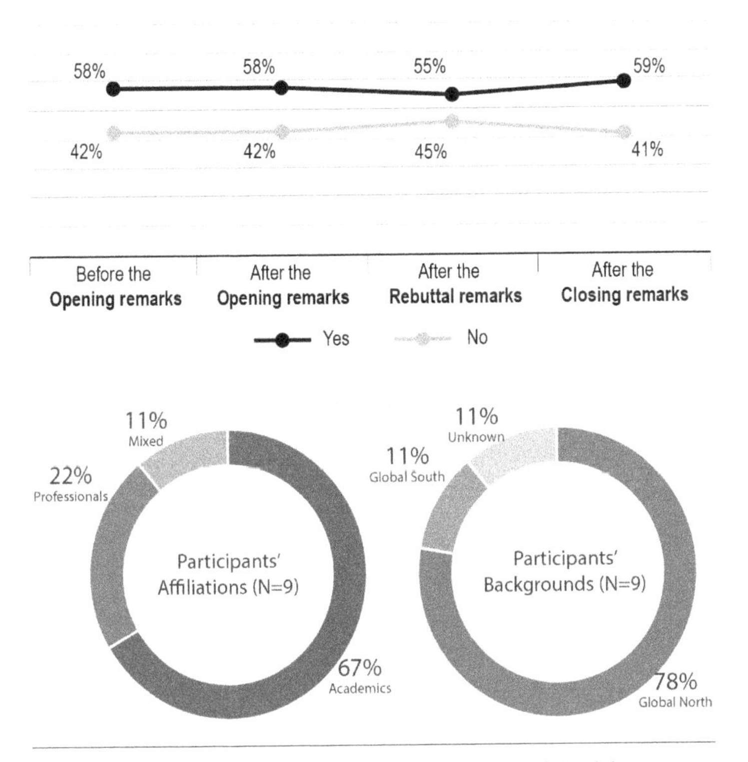

FIGURE 9.1 (*Top*) Results of votes at each stage of the debate. (*Bottom*) Distribution of affiliations and backgrounds among participants in the discussion.

when decision- and policymakers disapprove of specific lifestyles and ways of building. However, for many other experts, housing standards should address additional aspects, such as environmental protection, pollution reduction, and the protection of financial investments. But if the door leading to the implementation of housing standards in developing countries is already open, why stop here? Following this thought, some participants argued that standards should inevitably address other "key aspects" like neighborhood quality, crime control, and landscape design. Even the most committed defenders of standards seem to disagree on the most effective scope for regulations.

- *Secondary effects.* Participants shared the consensus that housing standards should not, in any case, penalize the poor and should instead aim at protecting them and their assets.
- *Target groups.* Participants did not seem to agree on the target of regulations and policy. Some believed that standards should only be imposed on builders and markets. Others argued that regulations should also apply to self-built constructions and informal settlements, notably during and after programs for upgrading slums. The control and monitoring of emerging and non-upgraded informal settlements remain a significant challenge.
- *Enforcement.* Neither the panelists nor the respondents specified how to enforce construction standards. However, they seemed to favor a stringent position toward builders, unscrupulous developers, and markets and a softer approach toward households responsible for building their own shelters or those who build through informal means. In some cases, participants argued that governments should act only as advisers for those who cannot afford regulated, formal housing units. The debate did not clarify whether central governments, municipalities, or both should be responsible for enforcing construction standards.
- *Scale.* Respondents shared different viewpoints regarding the scale of standards. For some, intervention should remain local and focus on communities, neighborhoods, and towns. Others suggested that general policies be applied nationally, perhaps motivated by concerns of urban fragmentation and inequality; these participants argued that bottom-up approaches could be replicated on a broader scale.
- *Time.* Timing seems to play a role in decisions to implement standards. Since residents are more aware of safety and security issues after disasters and calamities, such opportunities

provide ideal moments to implement new housing standards. A different problem emerges when considering more extended periods and development. Countries and their governing institutions reach maturity levels at other moments; the failed state of Haiti and a more prosperous Colombia are arguably at different stages of economic, institutional, and social development. Consequently, it is unclear at what stage housing standards should be embraced in the nonlinear economic development of countries.

- *Effectiveness.* We are aware that not all participants agree with us on this point, but sufficient evidence points to the consequences of not applying standards in developing countries, including but not limited to destruction, death, disasters, injuries, and health problems. Nonetheless, more research is still required to understand the specific benefits of applying and not applying standards, as well as the intended and unintended secondary effects of housing standards.

Surprisingly, a lack of knowledge about the efficiency and secondary effects of standards does not prevent participants from defending their application and enforcement. Defenders argue that politicians and technocrats should engage in drafting, approving, and enforcing low-cost housing regulations with taxpayers' money—a risky investment without a more thorough understanding of these impacts, benefits, and secondary effects.

Given the different layers of complexity regarding implementing standards for low-cost housing, this debate suggests a demanding policy agenda for developing countries that must also confront ethical considerations. The above-mentioned variables have profound implications for social justice, as they determine the success of housing standards in achieving better living conditions for the underprivileged. Standards can significantly shape settlements, as well as residential and land markets. If properly implemented, they can reduce risks and vulnerabilities and influence the building industry. But more importantly, this debate revealed much about how participants consider the least privileged and most vulnerable residents in their opinions on housing policy. We hope this discussion will profoundly affect decision-makers and politicians and produce a better housing agenda that addresses social justice issues. Finally, given our perspective summarized under "Effectiveness," we cannot stop believing that this debate also leaves scholars, students, think tanks, and international agencies with a significant research agenda, one we hope they will embrace for social justice.

NOTES

1. Michael Jacobs and Gelvin Stevenson, "Health and Housing: A Historical Examination of Alternative Perspectives," *International Journal of Health Services* 11, no. 1 (1981): 105–122; Jennifer Robinson, Allen J. Scott, and Peter J. Taylor, eds., *Working, Housing: Urbanizing: The International Year of Global Understanding-IYGU* (Cham, Switzerland: Springer Nature, 2016).
2. Uttam Kumar Roy and Madhumita Roy, "Space Standardisation of Low-Income Housing Units in India," *International Journal of Housing Markets and Analysis* 9, no. 1 (2016): 88–107.
3. Global Facility for Disaster Reduction and Recovery (GFDRR), *Building Regulation for Resilience: Managing Risks for Safer Cities* (Washington, DC: World Bank, 2015); World Bank, "From Urban Risk to Resilience: Building Safer Cities," 2016, https://www.worldbank.org/en/news/feature/2016/04/06/from-urban-risk-to-resilience---building-safer-cities; GFDRR, "Building Regulation for Resilience," YouTube, 2017, https://www.youtube.com/watch?v=cFEJJFsDpvI.
4. Colette Ghislaine Claudine Fransolet, "Universal Design for Low-Cost Housing in South Africa: An Exploratory Study of Emerging

Socio-Technical Issues" (MA thesis, Cape Peninsula University of Technology, 2015).

5. Tareef Hayat Khan, *Living with Transformation: Self-Built Housing in the City of Dhaka* (Heidelberg, Germany: Springer International, 2014).
6. A report by UN Habitat reports that there were an estimated 860 million slum dwellers in 2006; the number was expected to grow by six million a year through the first decade of the new century. WHO and UN-Habitat, *Global Report on Urban Health: Equitable, Healthier Cities for Sustainable Development* (Geneva: WHO Press, 2016), 169–173.
7. Kavitha Iyer, "Mumbai's Building Collapses and the Debris of Housing Policies," *Firstpost*, September 30, 2013, https://www.firstpost.com/mumbai/mumbais-building-collapses-and-the-debris-of-housing-policies-1141975.html.
8. WHO and UN Habitat, *Global Report on Urban Health.*
9. Diane Archer, "Baan Mankong Participatory Slum Upgrading in Bangkok, Thailand: Community Perceptions of Outcomes and Security of Tenure," *Habitat International* 36, no. 1 (2012): 178–184.
10. Edmundo Werna et al., "As Safe as Houses? Why Standards for Urban Development Matter," *Cities & Health* 6, no. 2 (2022): 404–417.
11. John C. Turner, "Barriers and Channels for Housing Development in Modernizing Countries," *Journal of the American Institute of Planners* 33, no. 3 (1967): 167–181; Alan Gilbert, "On the Mystery of Capital and the Myths of Hernando De Soto: What Difference Does Legal Title Make?" *International Development Planning Review* 24, no. 1 (2002): 1–20; Gonzalo Lizarralde, "Stakeholder Participation and Incremental Housing in Subsidized Housing Projects in Colombia and South Africa," *Habitat International* 35, no. 2 (2011): 175–187.
12. Jeroen Klink and Rosana Denaldi, "On Financialization and State Spatial Fixes in Brazil: A Geographical and Historical Interpretation of the Housing Program My House My Life," *Habitat International* 44 (2014): 220–226.
13. Brian C. Aldrich, "Winning Their Place in the City: Squatters in Southeast Asian Cities," *Habitat International* 53 (2016): 495–501.
14. GFDRR, *Building Regulation for Resilience.*

10

ON ENVIRONMENTAL PERFORMANCE INDICATORS

Should Construction and Urban Projects in Developing Countries Adopt Green Building Certifications Created in Developed and Industrialized Nations?

WITH JARED O. BLUM AND DAVID WACHSMUTH

Some scholars and practitioners have recently encouraged the adoption of green building certifications, such as LEED, BREEAM, EDGE, Passive House, and others, in developing countries.[1] Others have challenged the appropriateness of adopting certifications originally developed in and for industrialized economies in developing countries.[2] The former camp often notes that urbanization in developing countries is projected to rise from 46 percent in 2010 to 63 percent in 2050 and that the population density of these countries is expected to double over the next three decades. Therefore, the need to reduce the use of natural resources in such countries is urgent. Supporters of green building certifications also contend that they are increasingly adapting to local conditions in developing countries, thereby becoming valuable tools to:

1. Decrease negative impacts on nature and ecosystems, enhancing resilience

2. Reduce project and operation costs by encouraging the use of durable materials and efficient energy performance
3. Mitigate negative social impacts
4. Create awareness about environmental risks and stimulate the demand for sustainable building solutions[3]

Critics of adopting green certifications in developing countries often highlight how, by encouraging the adoption of standards formed for developed countries and economies, they neglect the local values, principles, and conditions in developing countries.[4] They note, for instance, the significant role played by informality, adaptation, and flexibility in less industrialized construction sectors. Critics also lament that certifications generally focus on buildings and neglect the complex and dynamic interactions among society, nature, cities, and territories.[5] They also argue that these certifications encourage the use of imported construction components and foreign technology, increasing project costs, carbon emissions, and dependence on industrialized solutions from the Global North. The prices of imported materials and technologies sometimes surpass the financial benefits of the energy savings achieved through green certifications. Finally, the prescriptive nature of green certifications and their emphasis on energy consumption often hinder local context-specific innovations and/or overlook social and physical vulnerabilities that might be of local importance.[6]

JARED O. BLUM: IN DEFENSE OF GREEN BUILDING CERTIFICATIONS FOR DEVELOPING COUNTRIES

Since Hammurabi's Code, societies have struggled with building performance standards. These efforts initially focused on structural integrity, fire safety, and—with the advent of electricity—mechanical performance. Today, several private-sector programs such as LEED and BREEAM rate and certify a building's "green" design attributes, which are assessed based on low energy and water consumption and recycled/recyclable materials, among other criteria. At the same time, the Passivhaus Institute and Passive House Institute US both focus on energy efficiency.

Green building standards preserve resources like water and energy, which are often in particularly short supply in developing countries. The U.S. National Institute of Buildings Sciences reports that green standards can result in energy, carbon, water, and waste reduction savings of 30–97 percent. The operating costs of green buildings can also be reduced by 8–9 percent, while property values can increase by up to 7.5 percent. Green certification programs have also produced other building benefits, such as higher productivity, greater climate resilience, and the increased health of occupants.

Why should these benefits be denied to developing countries? Given what we know about the toll of unsustainable energy and building materials on human health and the environment, it would be irresponsible to watch developing countries make the same mistakes. Buildings use about 40 percent of the world's energy; it is critical to make them more sustainable to reduce our environmental impact. Luckily, countries no longer must choose between economic development and sustainability. Sustainable building technologies and products have advanced tremendously in recent years, from solar panels to insulation, and costs have been reduced as a result. Building performance standards, rating systems, R&D, and pilot programs guide a more sustainable and economical way forward.

We recognize that no program covers all high-performance building criteria. Indeed, a building professional is likely to say that standards or certifications are context specific. Even in developed countries, building design and performance guidance or requirements are constantly evolving.

Developing countries experience the impacts of environmental challenges in building more keenly than anywhere else. Consider the case of Beijing, which almost shut down before the 2008 Olympics to reduce its appalling air pollution, or New Delhi, where schools closed and traffic was recently restricted due to critical air quality issues. Emerging economies must be able to take advantage of research, technology, and standard-setting processes developed in the Global North.

Though China called itself "a forever developing nation" when it opposed the Kyoto Climate Protocol in 1997, it now boasts the third most green buildings in the world. Energy and water use, occupant health, waste management, and site selection are considered as critical in China as in the European Union and the United States. India has already adopted the LEED India Certification program.

These cases do not imply that developing countries should necessarily adopt existing green certification programs in whole or in part, but they can and should use available knowledge generated by these programs. Developing countries should be able to choose building criteria and materials that best suit their unique circumstances, needs, and goals. Green standards do not discard old building strategies or materials because of age; the use of reflective roofs in Africa makes sense whether they are painted or made from EPDM, as new materials and technologies have not altered the physics of local building performance.[7]

National and subnational governments play a critical role in successfully adapting green standards such as BREEAM and LEED, as well as in promoting the regional diversity of traditions, cultures, materials, and skills. In developed countries, state and city governments have spearheaded rapid green building growth by adapting select aspects of green certification standards to their specific regions. Developing countries can similarly protect their local environments and human health by using local pilot programs based on programs such as Passivhaus and LEED. They can also encourage local advisory councils to set achievable higher performance standards, which will simultaneously accelerate economic progress. While imperfect and continually evolving, green building programs provide helpful guidance in developing countries.

DAVID WACHSMUTH: CHALLENGING GREEN BUILDING CERTIFICATIONS FOR DEVELOPING COUNTRIES

The need to increase the sustainability of the urban built environment quickly is one of the most urgent collective challenges human society faces. On one level, any good faith attempts to meet this challenge should be encouraged. Voluntary environmental programs such as LEED and BREEAM, which recognize developers who meet specific environmental criteria in their designs, indicate that the building industry increasingly values sustainability. However, many reasons to be skeptical of LEED and similar green building standards remain, particularly regarding their applicability to the developing world. First, the ability of green building standards to deliver sustainable benefits is likely overrated, even in wealthy countries. Second, the reasons to be skeptical of green building standards in the context of rich countries apply even more strongly to developing countries.

Why should green building standards be questioned even in wealthy contexts? First, no compelling evidence suggests that

LEED certification ensures more sustainable buildings; by contrast, research reveals that it may lead to *less* sustainable buildings.[8] Significant gaps remain in the LEED methodology, while building performance is frequently measured as worse than the projections that initially led to certification.[9]

Even if we accept that LEED certification is effective at reducing the environmental footprint of individual buildings, zooming out to a regional scale complicates the question. LEED standards are site-bound and will give developers credit for using "sustainable" materials even when the costs to transport these materials cancel out any resulting reduction in carbon emissions.[10] Some critics have suggested that the LEED criteria inadvertently encourage urban sprawl because they make site-specific demands, which are only feasible with large plot sizes.[11]

Moreover, the greater expense of LEED-certified buildings can have harmful secondary effects on regional land use that may swamp building-level sustainability gains. Since LEED-certified buildings tend to be more expensive and constructed in eco-conscious neighborhoods, poorer residents are pushed to less costly neighborhoods located in urban peripheries, where carbon footprints are often recorded as higher. As David Wachsmuth and colleagues conclude, adopting a regional perspective on costly building or local environmental improvements reveals that "many sustainability gains are simply a regressive redistribution of amenities across places."[12]

Concerns about the effectiveness of green building standards are especially relevant in developing countries, where the cost increases of green building are proportionately higher and tend to fall on more vulnerable populations. Urban sustainability outcomes can be improved in a more cost-effective and socially equitable fashion by prioritizing investments in dense, affordable housing and public transit; such policies distribute environmental gains more broadly among the urban population instead of concentrating them in a few expensive buildings.[13]

Finally, even if we grant that green building standards are effective and worth the cost in developing countries, the simple fact remains that construction in developing countries should follow local standards for green buildings. By contrast, Cole and Valdebenito found that 80 percent of the 136 countries where LEED or BREEAM are active do not have any domestic building standards system in place; they rely exclusively on one of these two international standards for green buildings.[14] Yet LEED, in particular, is relatively inflexible to diverse local contexts, whether cultural or environmental.[15] Different places have different sustainability needs, and these should be reflected in building standards if the latter are to be practical rather than a marketing tool for new high-prestige urban developments.

In sum, while building greener cities is an urgent task, certifying greener buildings with LEED or BREEAM does not seem to provide the most socially just or cost-effective way to accomplish that task in developing countries.

PARTICIPANT'S COMMENTS

Faten Kikano, PhD candidate, Faculty of Environmental Design, Université de Montréal, Canada

The twentieth century witnessed a return to vernacular and traditional approaches to building. Hassan Fathy was one of the pioneers of sustainable architecture for and by the poor, integrating traditional materials with modern architectural principles, as well as involving the less fortunate in the design and building process. His approach helped reduce labor costs, created jobs, and involved communities, in the process fortifying a sense of place attachment. Salma Samar Damluji was another architect who partly worked with Fathy. She led research on earth architecture and building technology to address social, environmental, and architectural overlaps in urban and rural fabrics. Focused on construction after disasters, Rohit Jigyasu was one of many researchers who demonstrated that local knowledge and capacity in building techniques and materials can reduce vulnerabilities, and that vernacular architecture can prove more resilient than contemporary methods.

Emphasizing the value of traditional architectural knowledge and building techniques contradicts the widespread generalization that people in developing countries tend to have a less sustainable lifestyle.

In conclusion, while certifications concern buildings, they also involve economic and sociocultural factors and may, therefore, produce serious drawbacks for vulnerable populations. Moreover, climate, landscape, nature, culture, and society differ from one place to another, and certifications of any kind can be adopted only if they do not disregard the specific context where they are implemented.

JARED O. BLUM'S REBUTTAL REMARKS

It is no coincidence that green building certification programs such as LEED and BREEAM originated contemporaneously with two benchmark international environmental agreements: the 1987 Montreal Protocol and the 1992 United Nations Earth

Summit in Rio. The former deal concerned ozone depletion, and the latter created the United Nations Framework Convention on Climate Change (UNFCCC). These landmark international agreements were formulated to address present urgencies, and both anticipated using technology transfers to attain their critical global goals.

Skepticism about whether developed countries can create transferable standards for developing countries to use in the built environment is understandable. Concerns about material costs, cultural dissonance, and availability of technology, technical expertise, and skilled labor can be raised. However, the perfect should not be the enemy of the good.

The Montreal Protocol—the most successful international environmental agreement in history—demonstrated that many of these impediments could be overcome. With the creation of a multilateral fund comprising seven industrialized nations and seven developing countries, ozone-depleting chemicals are being phased out globally, partly through technical assistance, training, and capacity building in developing countries. In November 2017, the UN Environment Programme announced that, under the Montreal Protocol, developing nations would receive more than $540 million from 2018 to 2020 to continue phasing out harmful chemicals.

Why is a similar effort not put forward in developing countries for green buildings? By following the Paris Agreement, developing countries have signed on to limit global temperature rise to less than 2°C; countries from India to Ecuador are taking measures to fulfill this commitment. Since nearly two-thirds of the world's countries do not have any building energy codes, using comprehensive and voluntary standards or rating programs could help these nations address additional critical issues such as water quality and efficiency, safe indoor air quality, and waste reduction. Against the opinion that implementing green initiatives would be too expensive and challenging for developing countries, I say that bold action to use resources more efficiently will ultimately serve more people and cost less.

The Green Climate Fund was established in 2010 through the UNFCCC to help countries use new technologies to upgrade building stock. Indeed, the International Energy Agency recommends offering developing countries access to global finance to help them adopt best practices and high-performance technologies into their building traditions. Armed with new knowledge and a growing supply chain, developing countries could avoid hurdles and mistakes that industrialized countries are still struggling to overcome, including their overreliance on fossil fuel energy and overengineered buildings that consume too many resources.

Some observers view the adoption of a developed country standard such as LEED in emerging economies as a form of neocolonialism or as insensitive to the possible socioeconomic impacts of new approaches to community living. While these concerns are certainly valid, localized innovation can address specific contexts. For example, the World Bank has developed EDGE, a new application focused on emerging markets, which uses some of the LEED categories in a simplified compliance path.

A partnership of local officials, building designers, contractors, and manufacturers dedicated to green strategies is evolving in developing countries. Such partnerships will be needed to create community resilience in developed and developing countries as we confront rapidly changing climate conditions. With resilience planning, natural hazards like floods, droughts, wind, and fires do not have to become natural disasters. USGBC has already recognized the global need for resilience by incorporating RELi, a resilient design platform, into the LEED rating system.[16] The future impact of climate change on our communities is a challenge for developed and developing countries alike; a globally recognized, consensus-based process may address the potential effects of climate change and provide all countries with the ability to reap the rewards of sustainable and resilient design and construction.

DAVID WACHSMUTH'S REBUTTAL REMARKS

Blum and I agree that reducing energy consumption in (and, more broadly, decarbonizing) cities and societies is an urgent task. LEED and BREEAM are voluntary programs; developers can adopt or avoid them as they see fit. The fact that a large and increasing number of construction projects in the developing world—albeit primarily those built by multinational, rather than local, developers—are working to achieve LEED gold or similar certification demonstrates the economic value of these programs. If their monetary value—which consists partially in saving from lower energy and water use and partially in increasing the prestige of a development for environmentally conscious buyers or tenants—motivates some environmental action, then they should be applauded.

However, at least from the perspective of fighting climate change, evidence regarding adopting green building standards remains mixed. While Blum broadly discusses sustainability in terms of environmental and economic considerations, I believe that we should evaluate the suitability of green building standards or any sustainability related policy with respect to its potential to reduce human-caused greenhouse gas emissions, as opposed to other important considerations. Though many studies have found that LEED-certified buildings have lower carbon footprints than non-LEED buildings, other studies conclude the opposite.[17]

It is possible that green building standards could be designed to reduce GHG emissions wherever they are implemented. Thus, the question as to whether they should be implemented concerns opportunity costs. Given other possible strategies for reducing urban carbon footprints, is it sensible for developing countries to devote scarce financial resources and attention to green schemes? Since time and money are limited, governments and societies need to choose how and where to spend both.

Blum makes the vital point that the impacts of unsustainability are felt most strongly in the developing world. This fact implies an ethical responsibility for developed nations to mitigate climate impacts. Building standards will not help the Global North achieve this mission because the subordinate environmental position of developing countries has little to do with buildings and much to do with how the global economy operates. Our world is marked by a spatial division of labor: high-value, low-pollution economic activities occur in rich countries; low-value, high-pollution economic activities occur in developing countries. Poor people tend to live in areas more vulnerable to sea level rise and other climate-related risks, while rich people and economically prosperous countries can better afford to protect themselves from these risks.

Referencing a set of benefits ascribed to green building standards, Blum asks, "Why should these benefits be denied to developing countries?" We might ask the same question about the range of social and material divergences between the developed and the developing world, which have led to worse environmental outcomes and have placed greater pressure on the natural environment in poorer countries. Green building standards will not solve these, but we might begin to address them with equity-focused urban environmental policy—and that is where I think we should invest our scarce time and money.

PARTICIPANT'S COMMENTS

Julien Deschênes, graduate student, Université de Montréal, Canada

It is worth emphasizing or developing further a few points raised in this debate. As Blum states, green certifications tend to reduce operating costs and raise property values. Given those benefits, implementation of certifications appears to be a good fit for developing countries, as residents usually have lower salaries and higher expenses for utilities like water or electricity. Green building certifications would therefore have a greater impact in alleviating household expenses in developing countries. However, access to materials and proper construction methods appears too costly to offset eventual operational benefits for smaller households. Finally, I have always found green certifications to be quite rigid, which I think blocks creativity and innovation by creating practical norms that appear virtuous and almost beyond questioning. To the public, green certifications are perceived as holy symbols of sustainable development and often as the sole approach to green construction or planning.

JARED O. BLUM'S CLOSING REMARKS

In my closing remarks on this interesting debate, I first want to compliment my colleague Dr. David Wachsmuth. His scholarly research and reasoned prose in dissecting the possible drawbacks of applying green building standards make me optimistic that sustainability practitioners, local community developers, and public officials can reach a consensus on equitable action.

My initial thesis in this debate still stands. The robust process used to create LEED, BREEAM, Passivhaus, and other green-certified standards in developed countries provides a transparent paradigm for developing countries. No profession, industry, or type of public official controls this process, whether in the United States, the European Union, Australia, or anywhere else. Consequently, the building science used to develop the recommended certifications has validity in the deserts of the American Southwest and the Sahara, from Bangkok to New Orleans.

Wachsmuth is concerned about the misuse of resources by developing communities that invest in green standards. I share his concern, so I have proposed that local assessments must determine the standards needed in specific contexts, particularly those that provide enough economic and social value to justify an initial investment in green certification.

It is not up for debate that developed and developing countries require shared actions to adapt to a climate future, for which existing building and community development practices are ill prepared. To this end, all tools should be welcomed.

DAVID WACHSMUTH'S CLOSING REMARKS

I want to touch on some of the points raised by the commenters who I think have managed to identify the issues at stake most clearly. Kent Thomas comments, "I'm from the Caribbean, and I think we are often financially unable to achieve the standards of BREEAM and LEED and can only do so for our greatest engineering feats (the most expensive ventures with a long design life)." This comment addresses one of my concerns with importing LEED in developing countries—that, in exchange for the certification, it allows high-prestige projects with large budgets an opportunity to make targeted efficiency improvements, but that these opportunities are far too scarce in poor, resource-constrained cities to be a sensible and general policy objective.

Faten Kikano critiqued the building-centric nature of LEED and BREEAM by noting that "while certifications apparently concern 'buildings,' they actually involve economic and socio-cultural dimensions, and may therefore represent serious drawbacks for vulnerable populations." We must consider how these standards implicate broader social relations.

Kirk Renaud offered a rebuttal to my critique and encouraged me to clarify my central argument: "David, I respect your concerns, but while you clearly agree that the need for global sustainability is urgent, you point out some minor flaws of certification programs without offering any alternatives. Given the need to avoid wasting precious time while the planet heats up, aren't we better to take a few less-than-perfect steps forward than doing nothing while Rome burns?"

However, I think I have offered an alternative to the certification programs by suggesting we focus on equity-focused urban sustainability interventions. I believe these will achieve better sustainability outcomes and be more acutely responsive to the diverse economic and social contexts of large developing-world cities. I pointed to investments in public transit and affordable housing as solutions that can be broadly applicable across cities in the developing world. Still, others may be worth prioritizing in specific contexts.

Mauro Cossu, another participant, offered a much more precise articulation of this position in his comment than I managed to do in either of my interventions:

> The problem for the urban poor is not access to sustainable housing, but their ability to maintain social networks and

> connections to places offering economic opportunities. They have other priorities: access to work, health care, and education. So, the location of affordable houses is fundamental, if not more important than the quality of the houses. Integrated urban solutions should not focus on green housing but also—and above all—on land and transport policies. For instance, walkable cities are more sustainable than those characterized by land consumption and sprawl because of a car-dependent infrastructure. To conclude, a city is not a collection of houses, and an increase in the number of green housing units does not measure urban sustainability. We should instead measure how real "capabilities" (Amartya Sen) determine and guarantee levels of urban accessibility and allow inhabitants to reach their places of interest—workplaces, social, and leisure sites.

This is a conclusion I can endorse.

PARTICIPANT'S COMMENTS

Mauro Cossu, PhD candidate, Faculty of Environmental Design, Université de Montréal, Canada

I certainly agree with the authors and other participants in recognizing a problem in the ever-increasing pressure humans place on Earth's ecosystems. I also see an opportunity to intervene locally with structural interventions in the urban built environment. Local is global! If our goal is to reduce the impact of human activities, the adoption of green building certifications in developing countries is not the most appropriate, effective, or ethical solution. The environmental crisis is a global concern faced by international organizations and nation states, but "solutions" are often conceived and applied with respect to single cities. Many weaknesses and contradictions emerge in addressing environmental crises. If major environmental challenges play out at the city level, then solutions should be based on integrated policies that allow us to redefine and broaden the spatial and social terms of urban sustainability.

THE MODERATOR'S CLOSING REMARKS

Green building certification systems are imperfect solutions to contemporary environmental challenges, but this should not discourage professionals and decision-makers from adopting them in developing countries.

Participants in this debate consider that the most essential action in a climate-changing world is to "begin the journey" toward sustainability awareness and environmental protection. Given this belief, the debate quickly focused on a significant ethical dilemma: whether implementing imperfect solutions suffices or whether developing countries should strive for higher objectives and more complex and challenging strategies to achieve them. This dilemma has profound ethical implications and raises crucial questions as to whether green solutions are "the perfect is the enemy of the good" in environmental protection and the fight against climate change. How just and reasonable must individual intentions and actions be in a world constrained by economic, political, cognitive, and administrative barriers? Given the increasing difficulty of this journey, is a modest step

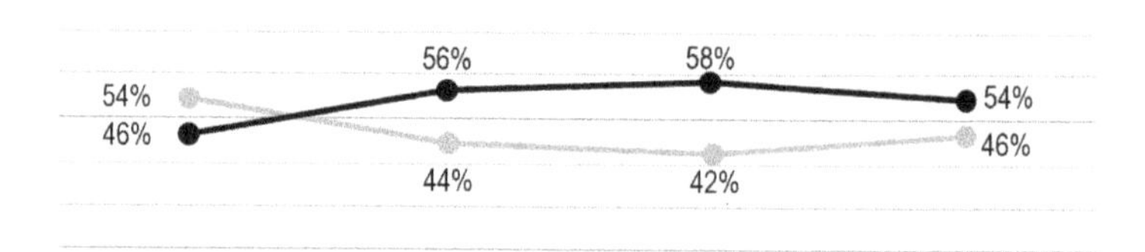

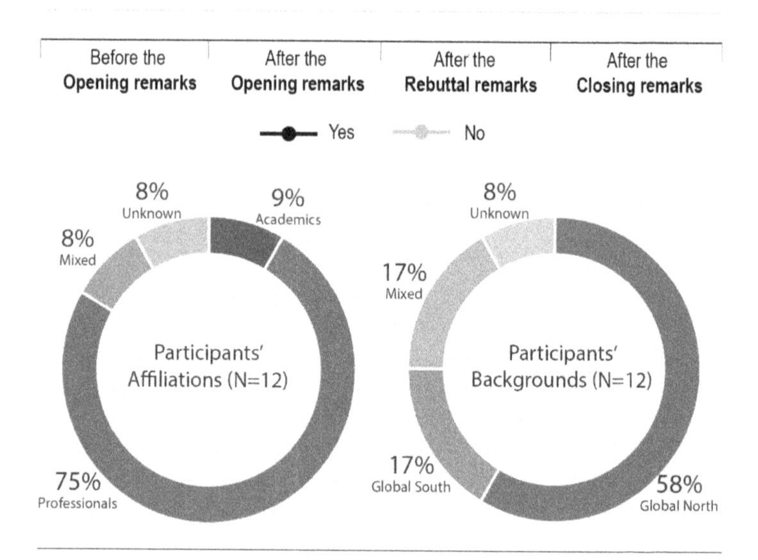

FIGURE 10.1 (*Top*) Results of votes at each stage of the debate. (*Bottom*) Distribution of affiliations and backgrounds among participants in the discussion.

toward a more sustainable world morally worthwhile? Does engagement hold an intrinsic value?

This debate was tight: whereas about 55 percent of voters defended the adoption of green building certification systems, roughly 45 percent raised serious doubts regarding their utility. Given the increasing acceptance of LEED, BREEAM, and other international certifications in global building standards, it is unsurprising that participants emphasized the positive outcomes of standardizing environmental principles across developed and developing countries. A rapid shift marked the distribution of votes. After the first round of the debate, nearly 55 percent of participants consistently defended adopting green building certification systems in developing countries. Some participants identified common drawbacks, limits, and side effects of green building certification systems designed in the Global North, but the effectiveness and value of certifications remained unquestionable. Most participants considered that green building certifications help to raise awareness about environmental issues and thus play a fundamental role in the journey toward sustainability.

Jared O. Blum, an expert in environmental issues in the United States, believes that the challenges of implementing green building certification systems can be, and are already being, addressed. David Wachsmuth, an urban planning professor at McGill University, questions the pertinence of these certification systems and their capacity to produce long-term and wide-scale changes. In a world with sufficient resources, any green solution, regardless of its scope, would probably be welcome. But in a context of scarce resources and limited public and political attention to environmental issues, Wachsmuth argues that we cannot afford to adopt green certification in developing countries. Instead, developing countries should focus on socially oriented interventions in domains such as affordable housing and transportation.

How do green certification systems hold in the face of climate change, disasters, and risks? Blum argues that such considerations have been considered, for example, in a resilience design platform called RELi, which the LEED rating system now includes. Blum and some participants believe that there is no need for developing countries to "reinvent their wheels." They see an opportunity for developing countries to adapt existing technologies and construction strategies, avoiding development stages characterized by pollution and environmental degradation. Wachsmuth and other commentators remind us of the ethical responsibility to prevent environmental damage in developed nations, which most strongly implicates poor communities. Wachsmuth and some participants seek local solutions in an industry and a construction sector shaped by informal solutions, urban inequalities, and informal businesses. Wachsmuth concludes, "These problems aren't going to be solved with green building standards, but we might begin to solve them with equity-focused urban environmental policy."

Some questions remain unresolved by the debate; some probably require more research and evidence. Are developed and developing countries fighting the same environmental problems? Do certifications produce the same effects in developing and developed nations? Do developing countries or developed ones have a greater impact on the environment? Regardless of their positions or answers, all participants in this debate seem to believe that we have a significant ethical responsibility toward society and the environment.

NOTES

1. LEED (Leadership in Energy and Environmental Design) was developed by the nonprofit U.S. Green Building Council (USGBC); EDGE (Excellence in Design for Greater Efficiencies) was developed by the IFC (International Finance Corporation) to facilitate an easy and more accessible certification process; BREEAM (Building Research Establishment Environmental Assessment Method) was launched by the Building Research Establishment (BRE) in the UK; Passive House (*Passivhaus* in German) was developed and widely

practiced in Germany and Scandinavia. See Amos Darko and Albert Ping Chuen Chan, "Strategies to Promote Green Building Technologies Adoption in Developing Countries: The Case of Ghana," *Building and Environment* 130 (2018): 74–84; Varun Potbhare, Matt Syal, and Sinem Korkmaz, "Adoption of Green Building Guidelines in Developing Countries Based on U.S. and India Experiences," *Journal of Green Building* 4, no. 2 (2009): 158–174.

2. Albert Ping Chuen Chan et al., "Critical Barriers to Green Building Technologies Adoption in Developing Countries: The Case of Ghana," *Journal of Cleaner Production* 172 (2018): 1067–1079.
3. Yanan Li et al., "Green Building in China: Needs Great Promotion," *Sustainable Cities and Society* 11 (2014): 1–6.
4. Felix Kin Peng Hui et al., "Green Buildings in Makassar, Indonesia," in *Green Building in Developing Countries*, ed. Zhonghua Gou (New York: Springer, 2020), 109–127.
5. Hong-Trang Nguyen and Matthew Gray, "A Review on Green Building in Vietnam," *Procedia Engineering* 142 (2016): 314–321.
6. Zhonghua Gou and Stephen Siu-Yu Lau, "Contextualizing Green Building Rating Systems: Case Study of Hong Kong," *Habitat International* 44 (2014): 282–289.
7. EPDM (ethylene propylene diene terpolymer), a resilient synthetic rubber membrane, saves heating and cooling costs.
8. John H. Scofield, "Efficacy of LEED-Certification in Reducing Energy Consumption and Greenhouse Gas Emission for Large New York City Office Buildings," *Energy and Buildings* 67 (2013): 517–524.
9. Jeroen Van der Heijden, "On the Potential of Voluntary Environmental Programmes for the Built Environment: A Critical Analysis of LEED," *Journal of Housing and the Built Environment* 30, no. 4 (2015): 553–567; Monica Ponce de Leon, "Constructing Green: Challenging Conventional Building Practices," *Constructing Green: The Social Structures of Sustainability*, ed. Rebecca L. Henn and Andrew J. Hoffman (Cambridge, MA: MIT Press, 2013), 333–340; Andrea Parisi Kern et al., "Energy and Water Consumption During the Post-Occupancy Phase and the Users' Perception of a Commercial Building Certified by Leadership in Energy and Environmental Design (LEED)," *Journal of Cleaner Production* 133 (2016): 826–834.
10. Ponce de Leon, "Constructing Green."

11. Robert Orr, "The Problems with LEED," The Project for LEAN Urbanism, 2014, http://leanurbanism.org/wp-content/uploads/2014/06/Orr-LEED.pdf.
12. David Wachsmuth, Daniel Aldana Cohen, and Hillary Angelo, "Expand the Frontiers of Urban Sustainability," *Nature* 536, no. 7617 (2016): 391–393, 392.
13. Chen Cohen, David Pearlmutter, and Moshe Schwartz, "A Game Theory–Based Assessment of the Implementation of Green Building in Israel," *Building and Environment* 125 (2017): 122–128.
14. Raymond J. Cole and Maria Jose Valdebenito, "The Importation of Building Environmental Certification Systems: International Usages of BREEAM and LEED," *Building Research & Information* 41, no. 6 (2013): 662–676.
15. Ozge Suzer, "A Comparative Review of Environmental Concern Prioritization: LEED vs Other Major Certification Systems," *Journal of Environmental Management* 154 (2015): 266–283.
16. RELi uses a holistic approach to resilient design to support rating systems and leadership standards; https://www.gbci.org/reli.
17. Ali Amiri, Juudit Ottelin, and Jaana Sorvari, "Are LEED-Certified Buildings Energy-Efficient in Practice?," *Sustainability* 11, no. 6 (2019): 1672; Omair Awadh, "Sustainability and Green Building Rating Systems: LEED, BREEAM, GSAS and Estidama Critical Analysis," *Journal of Building Engineering* 11 (2017): 25–29.

11

ON ADAPTATION

Is Adapting to Climate Change Our Best Choice?

WITH DEBORAH HARFORD AND SILJA KLEPP

Since the 1980s, scholars have debated whether decision-makers should focus on reducing greenhouse gas emissions (mitigation), reducing the risks posed by climate change (adaptation), or both. At first, mitigation prevailed in international climate policy, focusing on reducing carbon emissions. Early documents by the Intergovernmental Panel on Climate Change (IPCC) barely mentioned adaptation as a strategy to fight global warming. But in 2001, the IPCC's Third Assessment Report claimed that "adaptation is a necessary strategy at all scales to complement climate mitigation efforts."[1] Today, several experts accept that an "integrated portfolio" of mitigation and adaptation is required to confront climate risks.

In a 2007 article in *Nature*, experts argued that policymakers should overcome the "taboo on adaptation."[2] The authors stated that adaptation is necessary for three main reasons. First, even if CO_2 emissions are immediately drastically cut, carbon will remain in the atmosphere for decades; because of this "timescale mismatch," the effects of emissions on the climate will persist for years. Second, according to the "emissions fallacy," people suffer from several vulnerabilities unrelated to greenhouse gas emissions. Third, low-income countries and societies, which suffer more from the effects of global warming but are also less responsible for producing the emissions that cause it, will still require long-term risk reduction measures, what the authors term the "remediation" imperative.[3] Other defenders of adaptation have argued that individuals and communities have several "adaptive capacities" that can be relied upon and developed to deal with risks and disasters.[4] They consider human adaptation capacities an opportunity to "bounce forward" and prevent catastrophic events caused by hazards. Finally, urban experts often argue that adapting infrastructure and the built environment to risk is more environmentally and socially sound than replacing them with new constructions.[5]

Critics of adaptation often challenge its intrinsic value. Several scholars have argued that adaptation often focuses on technical solutions, failing to address the root causes of vulnerability, such as marginalization, exclusion, racism, colonialism, and other injustices.[6] They also argue that emphasizing policies that support physical adaptations to the environment has helped to "depoliticize" risk reduction and disaster response.[7] Disaster reduction should be considered a political rather than a technical issue that produces clear winners and losers and affects territories and societies in radical ways. Many scholars have found that "green infrastructure" has secondary risk effects, such as gentrification and displacement. Maladaptation perpetuates unsustainable development patterns and exacerbates inequality and environmental degradation.[8] Critics of adaptation wonder who should pay for adaptation measures, which rarely benefit poor and marginalized individuals and communities. They also lament that the concept of "adaptation capacity" is often deployed as a framework to transfer responsibility for risk management to individuals and the private sector. At best, adaptation perpetuates a neoliberal conception of risk reduction, and at worst, it contributes to disaster capitalism.[9] For such critics, the discourse of effective adaptation is dangerous because it encourages industries and political elites to maintain current emissions and pollution levels. Furthermore, by promoting a shared approach to climate response that thrusts responsibility for risk reduction upon individuals and communities who are (or must become) "adaptable," political and economic elites dilute their accountability regarding pollution, disaster risk creation, and environmental degradation.

DEBORAH HARFORD: ADAPTING TO CLIMATE CHANGE IS OUR BEST CHOICE

Adaptation to climate change is an urgent global priority for individuals, communities, ecosystems, and economies coping with natural disasters and their effects on health, equity, infrastructure, biodiversity, and socioeconomic systems.

A root cause of the global spread of viruses such as COVID-19 is habitat loss due to carbon-intensive developments, which infringes on ecosystems and displaces species already at risk because of climate change. The multiscale disruptions caused by this pandemic will not compare to the global challenges for water, food, health and well-being, energy and economic security, and geopolitical stability that climate change will cause over the coming decades. Slower onset impacts, such as sea level rise and melting glaciers, will devastate coastal and delta communities and freshwater supplies. The biggest concerns faced by human populations will include impacts on the most vulnerable individuals and communities, as well as their geographic displacement.[10]

Even if we reduce carbon emissions, the impacts of climate change will increase between now and 2050. At that point, the results of our efforts—or lack thereof—to mitigate climate change will become more visible, with high emissions trajectories threatening runaway climate change. Adaptation, therefore, cannot be separated from emissions reduction; the two should be planned in an integrated manner I refer to as low-carbon resilience. Likewise, adaptation cannot be separated from research, planning, and action, which are needed to develop knowledge of how it can work.

The central purpose of adaptation is to embed strategic responses to climate change risks into policy, planning, and actions—a process known as mainstreaming—to reduce vulnerability and build social resilience. This approach to risk management is essential if we are to minimize climate change's negative effects on global health, economies, infrastructure, ecosystems, and community well-being. Effective adaptation acknowledges that addressing poverty, health, and equity is critical to building resilience and supports the UN Sustainable Development Goals, the Sendai and Paris Agreements, and the Aichi biodiversity goals.

Luckily, humans are extraordinarily adaptive beings, and we have solid climate science at our disposal to support place-based and systemic risk responses based on community values and priorities. These responses include both "incremental" approaches that accommodate and reduce impacts and "transformative" adaptation that reimagines complex systems.[11] In one inspiring example that integrates both methods, Bangladeshi experts are leading their country's response to sea level rise by helping coastal farmers transition to salt-tolerant crops. At the same time, they are reimagining possibilities for future generations by preparing children to train in different jobs and upgrading inland cities to accommodate mass relocation.[12]

Adaptation is rapidly evolving as we learn more about what works and what needs to improve in our approach to climate action. Emerging focus areas include cocreating solutions with Indigenous communities, infrastructure retrofits, managed retreats, understanding gender impacts and mental health needs of individuals, nature-based solutions, and more.

Collaborations dedicated to developing such solutions are accelerating at all levels of society worldwide. Climate impacts will constantly test what we put in place now, from policy to buildings. If we work together to build widespread low-carbon resilience, we can strategically improve various social and economic priorities.

SILJA KLEPP: ADAPTING TO CLIMATE CHANGE IS NOT (NECESSARILY) OUR BEST CHOICE

During my research in Kiribati, I saw a signboard in the main office of the Kiribati Adaptation Program, financed by the World Bank and other international donors, that said: "Adapt or Perish."[13] This imperious statement symbolizes how climate change adaptation has become a new "imperative" for the Global South. Narratives of climate change adaptation are increasingly urgent, and the need for a certain kind of "expert" knowledge to realize "successful" and "effective" adaptation; otherwise, populations will be "doomed."

As explained in the moderator's opening remarks, the shift in climate change governance from mitigation to adaptation is problematic. In the 2000s, scholars and decision-makers envisioned adaptation as an alternative policy to insufficient mitigation efforts. It also became an area rich in business opportunities for international organizations, consultants, and NGOs.

Western technical-fix projects similar to those highly criticized in the 1980s and 1990s dominate development aid today.

In my fieldwork interviews, I often heard citizens complain that consultants did not listen to the I-Kiribati (the people of Kiribati). Consultants instead followed their own frameworks and cultural practices, often in response to profit-oriented objectives. Most climate adaptation consultants are "ticking their boxes," as noted by one of my research partners in Kiribati. They create knowledge and practices that must fit into tight schedules (one to two years), produce specific outcomes, and be marketable.

The "adaptation" industry currently thriving is one that rarely considers social contexts, cultures, and power relations in its decisions. In this industry, non-Western ontologies considered far from the logic of mainstream adaptation efforts are ignored, especially in international cooperation activities. Today, entire development programs have been reframed to focus on technology-driven adaptation measures because climate adaptation offers a more profitable market than dealing with other pressing problems, such as poverty, food insecurity, and domestic violence. Many scientists expect to obtain measurable, even if only modest, results. This goal detracts us from pursuing ambitious yet intangible outcomes, like developing a more holistic understanding of resilient communities.

Several stakeholders are trying to work differently by applying new methods and frameworks to adaptation. For example, Daniel Morchain of the International Institute of Sustainable Development (IISD) is experimenting with theater play to empower vulnerable individuals and encourage more holistic and emancipatory approaches.[14] According to Morchain, "adaptation must be transformative, or else it is dangerous."

I agree with Morchain. Adaptation can be dangerous if it does not contribute to urgent socioecological transformations in the field of disaster risk reduction that address the failures of Western lifestyles. Otherwise, adaptation will only perpetuate current institutional priorities and the status quo. So long as climate change adaptation interventions are drafted in the headquarters of Western organizations without consideration of cultural specificities or power relationships within and between communities, the risk of inflicting violence and injustices on the most vulnerable individuals and populations remains high. We need a new way of thinking and acting rooted in solidarity, connectivity, and transformative change—not only within a nation-state system but as a community of human and nonhuman beings connected in myriad ways in the Anthropocene.

PARTICIPANT'S COMMENTS

Oleg Zurmuehlen, graduate student in human geography, University of Bonn, Germany

Political negotiation takes time, and humans are already lacking time in our efforts to adapt to climate change. This lack of time has arisen because we are stuck in fixed ways of thinking and acting, especially in terms of economics. However, I do not think that anyone arguing for the politicization of adaptation strategies is throwing the baby out with the bathwater. At best, politicization works toward long-term efficiency of climate change adaptation. It starts with the process of choosing which technology to use for adaptation. Politicization is inextricably entwined with worries about technoscientific issues of adaptation. I agree that adaptation is a sociotechnical problem, but I disagree with proposals to separate politics and technology in climate change interventions—technologies are political.

Should hydroelectric dams be built to ensure intensive irrigation in a drought-riddled monoculture landscape, thereby displacing local communities and destroying local ecosystems, or should monoculture landscapes be transformed by efforts to minimize the amount of water they consume, for example by shifting to regenerative agricultural technologies?

We need to change our modern definition of technologies to reconsider seemingly nonmodern ones like regenerative agriculture as sophisticated. The aim of politicizing adaptation includes folding nonmodern technologies into the set of technologies presented to decision-makers, as a diversity of choices feeds politics. Politicization encompass not only the business of politics as pursued by politicians but also involves debates and exchanging arguments like our present one, while also analyzing the power structures that frame our arguments. This process of politicization will inevitably inform sustainable long-term choices; it is a more responsible thing to do to prepare future generations for impending climate disasters.

DEBORAH HARFORD'S REBUTTAL REMARKS

Adaptation is gaining momentum globally because we have failed to stop climate change. Climate change is a problem facing everyone, but not equally; one of the most painful injustices resulting from our failure to curb climate change is that the lowest carbon emitters will suffer the worst ecological impacts. For instance, sea level rise resulting from heat accumulated in oceans means that small island states like Kiribati will face the disappearance of their homes and ancestral lands. While cutting emissions will slow rising tides, ecological destabilization is already occurring. The arduous process of developing responses to climate risks and impacts has not kept pace; as Klepp notes, such efforts add insult to injury if poorly designed.

We have seen similar issues affecting the Canadian Arctic, where climate change impacts are rapidly emerging and heavily affecting the Inuit native to the land. While Indigenous communities are innately resilient, colonial legacies have limited their adaptive capacity.[15] Moreover, many communities have voiced dissatisfaction regarding researchers from urban centers who enter and leave their communities for study purposes without providing them with tangible benefits.

Rather than refuting a need for adaptation, Klepp's opening statement makes an excellent case for reconsidering how it is conceptualized and actualized. In many ways, the critique she outlines is well understood. Numerous projects worldwide, including mainstreaming adaptation into other actions, reflect such progressive approaches. As the urgency to adapt to current circumstances grows, scholars and policymakers are addressing biases and flaws in funding, researching, planning, and implementing adaptation. They are also confronting a growing need to prioritize approaches to adaptation underpinned by equity, social justice, cocreation, and decolonization. The pressing contemporary need for adaptation may mobilize opportunities to rethink better working and living methods.

Developments in adaptation mirror other cultural advances, such as the global movement to recognize the UN Declaration of the Rights of Indigenous Peoples. For instance, Canada is engaged in a Truth and Reconciliation process to address the damages of colonization. In 2019, as part of a panel charged with developing criteria for Canada to measure progress on adaptation, the panel's Inuit representative and I coauthored a chapter titled "Translation of Indigenous Knowledge and Western Science Into Action."[16] Our report showcased the significant value Indigenous knowledge holds for adaptation. When respectfully codesigned with Indigenous experts, funding schemes, policy, and projects can benefit from the application of Western science and research. We identified the need to support this collaborative approach to adaptation by investing in building awareness, capacity, relationships, and support for knowledge, as well as through local solutions developed by and for those experiencing climate-induced changes to their environments.

The COVID-19 pandemic highlighted inequalities in the global system in ways that hold significant lessons for determining the next steps in climate action. It is more necessary than ever that we work respectfully together to ensure that social justice and equity underpin adaptation.

SILJA KLEPP'S REBUTTAL REMARKS

Given the opening remarks written by my "opponent," Deborah Harford, and inspiring comments by participants, many of us agree that climate change adaptation involves problematic knowledge-power relations that often hinder fair and transparent outcomes. So, if we all agree, why is it so hard to implement fairer climate change adaptation interventions?

Climate change adaptation is a social justice problem, not just an environmental one. This means that the road to change begins with thinking differently about human-environment relations. Since the so-called Enlightenment, we have lost our capacity to think of human beings as a fundamental part of nature. Fortunately, in recent decades, literature across disciplines has significantly helped us to conceptualize naturecultures, socionatures—or the *Chthulucene* as theorized by Donna Haraway—in a much more inclusive and engaged manner.[17] I want to put forward the environmental justice perspective both as a way of rethinking climate change adaptation, questioning traditional and problematic knowledge-power relations, and as an approach to making climate change adaptation more transformative and just.

Environmental justice is rooted in environmental and civil rights activism. It has always been connected to Indigenous and First Nation groups' struggles to seek a fairer distribution of environmental consequences. Ontologies that help us to think of naturecultures more inclusively have played a crucial role from the beginning of the environmental justice movement in the 1980s. Since then, environmental justice has progressed in academia and among scholar-activists to embrace a threefold analytic perspective.

First, dimensions of distributive justice can help us consider those included and excluded by climate change adaptation measures in the "community of justice." Who will profit from adaptation money—the most vulnerable, the most affected, or the elites closest to the international aid community?

Second, procedural justice prompts us to question those who participated in the planning and realization of adaptation measures. Who has access to decision-making and forums that promote discussions on ethically just ways of adapting?

Third, justice as recognition, a concept introduced by David Schlosberg, closely connects distributive justice and procedural justice and further aims to recognize diverse ontologies and forms of knowledge.[18] Suppose we understand climate change adaptation as a complex set of narratives and practices. In that case, we might consider our most important tools to overcome the inequitable and imbalanced knowledge-power relations inherent to climate change adaptation. To reply to my dear "opponent," this approach demands that the coproduction of knowledge actively recognizes adaptation as a messy and challenging social process with a potentially open and transformative outcome.

I believe that systematically applying an environmental justice perspective illuminates injustices in adaptation more clearly and thus enables us to tackle them radically.

PARTICIPANT'S COMMENTS

Duvan Hernán López, PhD student, Sustainability at Polytechnic University of Catalonia (UPC), Spain

I think it is worth questioning the politics of adaptation, which has been an adjacent concern in this debate. The pressing need to solve technological challenges to optimize social metabolism cannot be doubted, but an evident asymmetry between the human capacity to promote technical innovations and the human capacity to articulate collective action on different scales limits this goal.

With modernity and globalization, humanity has successfully implemented a global administration of material and energy resources, but we still find it difficult to address conflicts among diverse subjects. Hence, it is necessary to develop awareness and understanding of intersubjective gaps and elaborate mechanisms to solve these conflicts.[a]

From this perspective, further reflections on adaptation are needed, such as "Resilience for whom?," "Who should pay?," and questions raised recurrently in this debate.[b] Beyond reflection, we must elaborate and implement mechanisms that guide adaptation and its narratives and resources to resolve broader social issues. I understand the environmental justice approach proposed by Silja and the urgency of championing biases and flaws in how adaptation is funded, researched, planned, and implemented, as recognized by Deborah.

I think we can achieve consensus in this debate but keep social considerations and priorities in mind as well.

[a] Meneses Hernán López Duván, "Los riesgos climáticos y el sentido humano: Claves frente a los conflictos ambientales desde el pensamiento de Teilhard de Chardin," *Razón y fe: Revista hispanoamericana de cultura* 284, no. 1453 (2021): 185–198.

[b] Susan L. Cutter, "Resilience to What? Resilience for Whom?" *Geographical Journal* 182, no. 2 (2016): 110–113; Sara Meerow, Joshua P. Newell, and Melissa Stults, "Defining Urban Resilience: A Review," *Landscape and Urban Planning* 147 (2016): 38–49; Daniel A. Farber, "Adapting to Climate Change: Who Should Pay," *Journal of Land Use & Environmental Law* 23, no. 1 (2007): 1–37.

DEBORAH HARFORD'S CLOSING REMARKS

Climate change is a social and political problem and must be addressed as such. Adaptation is one key element within a broader suite of necessary social transformations—equity and justice must be at the forefront of such consideration.

We have failed to reduce GHG emissions to the point of negligible impact. In the future, we will have to adapt to climate realities, either reactively through piecemeal responses that entrench conditions that create vulnerability or proactively alongside broader initiatives for social transformation.

To this end, adaptation raises a material and moral question: How can we stay loyal to social and environmental justice principles while working in a field located at the intersections of knowledge and power? Who defines adaptation, what adaptation measures should we deploy, and for whom?

Humans have the agency to create change in broader structures that limit our beliefs and behaviors. Our ability to understand the probable impacts of climate change provides us with what I call the "adaptive advantage": the opportunity to plan at the moment rather than be blindsided in the future, potentially alleviating suffering and hopefully initiating transformative changes that address multiple priorities.

Three priorities must inform adaptation to ensure that it does not merely provide a "technological fix," perpetuate the hegemony of the developed world, or operate according to a purely anthropogenic approach, but instead becomes a path toward genuinely sustainable development.

First, we must shift away from distinguishing adaptation and mitigation as separate approaches and move toward an integrated focus I refer to as low-carbon resilience.[19] For example, as flooding and heat-related natural events increase, it is more likely that large-scale, emissions-intensive gray infrastructure will become a default solution to these problems. We should not strive for this kind of response.

Second, climate change threatens countless species; ecosystem health underpins our survival. We must advance nature-based solutions as a dual mandate to protect humans and ecosystems from climate change and consider the protection of nature.

Third, if people remain vulnerable because of poverty, a lack of access to health care, and other social and political factors, our efforts at adaptation will fail. While climate change creates new problems, such as changing sea levels, it principally worsens such existing problems. A central priority of organizations and researchers working on adaptation must involve acknowledging social justice, equity, and the resources needed for effective responses. We must work to act as allies, building partnerships by supporting local communities, communities in the Global South, and Indigenous peoples to direct the structure and design of adaptation funding, principles, planning, and implementation.

If we prioritize these three considerations, we may develop transformative solutions based on principles that avoid unsustainable development scenarios and address the needs of the disadvantaged.[20]

SILJA KLEPP'S CLOSING REMARKS

One key takeaway from this debate is that it is crucial to understand climate change adaptation as a knowledge-power nexus at various spatial scales.

Deborah illustrated how she and her colleagues are helping heal the injuries of colonialization in Canada. Cooperating in postcolonial settings or settler communities requires examining their histories of violence perpetrated by colonial policies and practices. These policies have often destroyed communities, for example, through abusive education systems. The Canadian scholar Emilie Cameron showed how researchers' understandings of the terms "indigenous" and "local" could hinder political change and organize climate change adaptation measures in backward ways.[21] Enhancing community resilience in such cases often demands empowering respective communities.

In my closing remarks, I also want to draw attention to the vital role of science and scientists. It is beneficial that more scientists are acknowledging the political dimensions of climate change by making public statements or engaging as scholar-activists. We should demonstrate a similar critical approach and reflexivity in our public engagements as we do in our research, questioning the role played by our work in society and how we deal with its consequences.

PARTICIPANT'S COMMENTS

Steffen Lajoie, PhD candidate, Faculty of Environmental Design, Université de Montréal, Canada

My doctoral research began with a focus on increased investments in adaptation to climate change in *barrios populares* (slums) in Bogotá, Colombia during Gustavo Petro's mayoralty (2012–2015). Petro established adaptation as a driving principle to revise the Plan de Ordenamiento Territorial, the city master plan. He also redirected the municipal agency in charge of disaster prevention and response toward a mandate of climate change adaptation that integrated and funded pro-poor transformative methods such as participatory budgeting.[a] Soon after I began my research project, Petro was removed from office. Almost immediately thereafter, all projects associated with his tenure stopped and the bureaucrats linked to the project were replaced or removed. As a result, I shifted the geographic focus of my research to the Valle del Cauca department, with a focus on the cities of Cali and Yumbo. In these regions, adaptation also entered a tricky political muddle. Perhaps these imperfect or incomplete urban and infrastructural projects represent stages of change, adjustment, or adaptation—perhaps this is what transformation looks like in process.

In contrast to the Bogotá case, the flagship adaptation project in Cali funded by the National Adaptation Plan, "Plan Jarillon" (2016–present), is a mega-infrastructure project to address the aging dike along the river Cauca, which protects millions of inhabitants from regular and major flooding.[b] One might ask why authorities allowed so many people to live in a flood zone in the first place, and rightly so. Given the existing risk, any solution would be controversial. The present plan involved relocating about eight thousand families living informally on and near the dike. For some families, the project was an opportunity to have a new apartment in a different location, but for others it was a violation of rights that implicated the loss of their home and livelihood. Here "adaptation" was approached with good intentions from an ecological and public safety perspective. On the other hand, the plan continues to leave millions not included in the relocation plan living on the floodplain behind the dike. It also locks the neighboring city of Cali, with its population of five million, into a former pattern of attempting to control nature instead of addressing the sociopolitical causes of vulnerability. Further, the plan did not resolve the lack of "formal" housing and rural exodus that drove the most vulnerable to squat on the crumbling dike in the first place.

This example touches on three critical operational challenges: (1) working within the limits and realities of existing contexts, governance systems, and governments; (2) balancing the competing interests and priorities of community members and other actors; and (3) bringing social justice into practice. All challenges are inherently political and provide support for the argument calling for "transformative adaptation."

[a] Austin Zeiderman, "Adaptive Publics: Building Climate Constituencies in Bogotá," *Public Culture* 28, no. 2 (2016): 389–413.
[b] Rincón Carlos Eduardo Díaz, "Jarillón de Cali: ¿Cómo va su reconstrucción y el reasentamiento de las familias?," *El Espectador*, November 3, 2021, https://www.elespectador.com/colombia/cali/jarillon-de-cali-como-va-su-reconstruccion-y-el-reasentamiento-de-las-familias/.

In recent years, there has been a significant rise in attention to climate change models and related socioeconomic processes. For example, considerations of greenhouse gas emission projections like the representative concentration pathways (RCPs) now play a massive role in IPCC recommendations and general policymaking. Researchers identifying these pathways believe they provide "neutral" or "objective" guidance for policymaking, but this is untrue. Silke Beck and Martin Mahony have shown that decision-makers' reliance on RCPs has political implications.[22] It becomes harder for them to "think out of the box" or in a radically transformative way, as when politicians consider a narrow range of options that do not include alternative solutions.

I hope that we, as scientists, will not stop asking for better predictions or more positivist science. Yet I believe we need to emphasize the power relations that shape technical solutions and policy, such as UNFCCC negotiations. We need critical research on the unsustainable pathways that limit our work. Unfortunately, my impression is that we are heading in the wrong direction. Statistics, figures, and models seem so ontologically powerful that measurements and models dominate research questions on climate change and human-environment relations that would be better answered by interdisciplinary teams—teams that should include scholars from critical social sciences and the humanities. We are witnessing, for instance, the difficulty of social science research proposals to obtain grants in interdisciplinary funding calls. These observations should not lead us to inaction or disappointment but should instead remind us that we must work together in a courageous and connected way. We should not avoid the difficulties of inter- and transdisciplinary coproduction of knowledge but find more inclusive, creative, and reflexive ways to collaborate. Moreover, we should join in the political fight against climate change and generate more equitable ways of adapting to its impacts. The network for environmental justice EnJust is one such forum bringing together scholars, activists, and practitioners.[23]

PARTICIPANT'S COMMENTS

Tushar Pradhan, disaster risk management and climate change professional, Maharashtra, India

Adaptation is one of several ways humans can tackle the impacts of climate change and disasters, but it is only one among several solutions required. The current challenges we face (from carbon emissions to rising sea levels and diseases) are the result of poor governance and excessive exploitation of natural resources. Having a technical solution driven by scientific knowledge is indeed required—but in our quest for trying to resolve critical problems, we often forget the role played by each factor. There are huge inequalities in our societies that need to be resolved before attempting to adapt to new threats. I foresee adaptation as an opportunity to redress these gaps technically, socially, economically, and culturally. There is no one method that fixes all of these problems at once, but we do have new methods to resolve them individually. Adapting to the changing climate with sustainable strategies and diligently working on the SDG 17 goals and 169 targets can lead to achievable change.

THE MODERATOR'S CLOSING REMARKS: THE ETHICS AND POLITICS OF CLIMATE CHANGE ADAPTATION

Adaptation is still the best choice in our fight against climate change effects, but only if it is politically engaged, motivated by social justice, and aimed at reducing inequality and the root causes of vulnerability.

Approximately 63 percent of participants in this debate believe that adapting to climate change is the best choice for humanity moving forward. Yet the arguments espoused by researchers and respondents reveal that not just any adaptation is required or desirable. Adaptation focusing exclusively on technical solutions while failing to deal with the social, economic, and political conditions that lead to disasters is insufficient and may be dangerous.

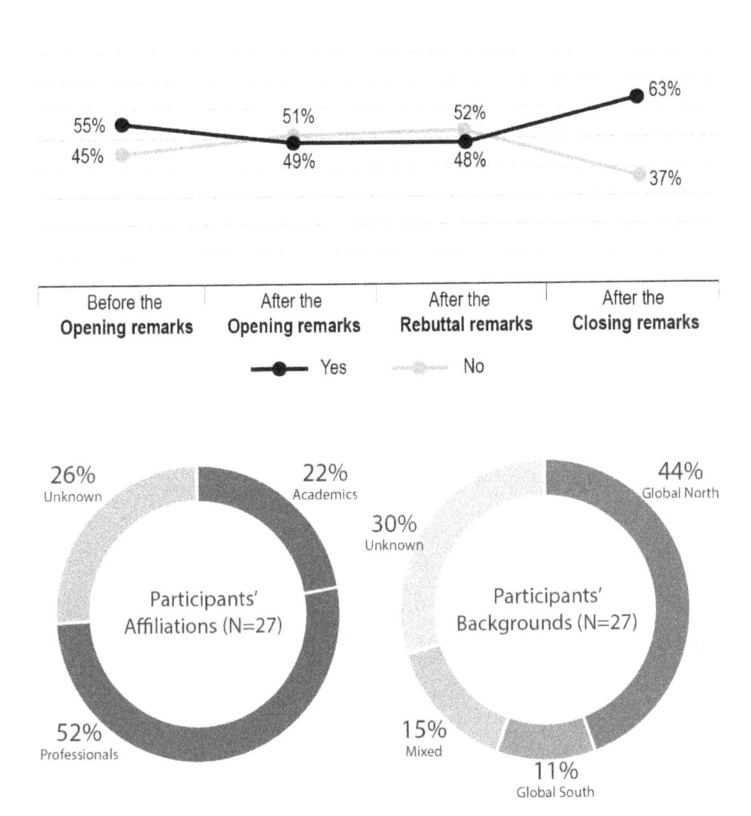

FIGURE 11.1 (*Top*) Results of votes at each stage of the debate. (*Bottom*) Distribution of affiliations and backgrounds among participants in the discussion.

Radical reductions of carbon emissions have proven difficult to achieve; indeed, were it not for major crises like the COVID-19 pandemic (see chapter 1, "On Fragility"), we might have considered this task almost impossible in the short term. We must therefore accept that some form of adaptation to a warming planet is necessary for our continued survival. Most participants agree that comprehensive strategies for disaster risk reduction must consider adaptation, but it should not detract from continuing efforts to reduce carbon emissions and the depletion of nature.

The adaptation needed today requires specific governance structures and must be ethically guided. Adaptation implies several capacities among local authorities, and it calls for institutions to recognize the interests, traditions, and expectations of the most vulnerable. Activists, decision-makers, professionals, and politicians working towards adaptation must identify power imbalances in their actions and policies and try to redress the harmful effects of colonialism, marginalization, exclusion, racism, and imposed segregation. Efforts to deal with natural hazards and environmental problems should not be separated from struggles for social justice. We must open spaces for dialogue in which we can recognize and address previous injustices. In other words, we must achieve what one of our participants called "negotiated resilience" (see chapter 3, "On Resilience"). In this way, adaptation can become a crucial vehicle for aligning efforts toward positive social, political, and environmental change in disaster risk reduction without leaving the poor behind.

Adaptation is no longer simply a way to describe how people deal with risk but has also become a normative approach to risk reduction adopted by international consultants, UN agencies, foreign charities, and other giants in the disaster industry. Adaptation efforts often focus solely on technical solutions (see chapter 9, "On Regulations," and chapter 10, "On Environmental Performance Indicators"), as defenders of adaptation tend to neglect its political implications—yet politics are always involved. Even though climate change is a global phenomenon, we must be aware that it does not affect all people and regions equally. Actions, including those taken for adaptation, have side effects and unintended consequences. Pursuing the wrong type of adaptation can perpetuate or even increase inequalities and power imbalances. We must ensure that the effects of adaptation do not increase vulnerabilities among the poor, minority groups, indigenous communities, and the historically excluded and marginalized.

The concept of adaptation must be "decolonized" if it is to be made more useful in reducing the risk of disasters and preventing the creation of new risks. Actions taken in adaptation initiatives can perpetuate poorer countries' and communities' dependence on wealthier ones (see chapter 2, "On Power Imbalances"). We must pay attention to who finances climate change adaptation (see chapter 7, "On Aid") and who decides where, for whom, why, how, and when adaptation is implemented (see chapter 6, "On Participation"). The type of adaptation required to deal with modern risks must recognize unique contexts and specific needs. It must look for alternative ways of producing culturally relevant interventions but should not exclusively regard local action as beneficial. Adaptation must also recognize that local elites and privileged groups may use it to create divisions, favor tribalism, facilitate local corruption, and oppress rival political groups. When implementing adaptation, we must consider local values and universal human rights.

This debate revealed the complexity of framing contemporary climate action. Participants' opinions changed over its two-week span. At first, 55 percent of participants believed adaptation was crucial (Yes). After a few days, the vote reversed, shifting slightly in favor of No. However, in the end, Yes prevailed with 63 percent.

NOTES

1. James McCarthy et al., eds., *Climate Change 2001: Working Group II: Impacts, Adaptation and Vulnerability, Third Assessment Report of the Intergovernmental Panel on Climate Change* (Cambridge: Cambridge University Press, 2001).

2. Roger Pielke et al., "Lifting the Taboo on Adaptation," *Nature* 445, no. 7128 (2007): 597–598.
3. IPCC, *Climate Change 2014: Impacts, Adaptation and Vulnerability*, Contribution to Working Group II, The Fourth Assessment Report of the Intergovernmental Panel on Climate Change (Cambridge: Cambridge University Press, 2014).
4. Barry Smit and Johanna Wandel, "Adaptation, Adaptive Capacity and Vulnerability," *Global Environmental Change* 16, no. 3 (2006): 282–292; Fran H. Norris et al., "Community Resilience as a Metaphor, Theory, Set of Capacities, and Strategy for Disaster Readiness," *American Journal of Community Psychology* 41, no. 1 (2008): 127–150.
5. Simin Davoudi et al., "Resilience: A Bridging Concept or a Dead End? / 'Reframing' Resilience: Challenges for Planning Theory and Practice / Interacting Traps: Resilience Assessment of a Pasture Management System in Northern Afghanistan / Urban Resilience: What Does It Mean in Planning Practice? / Resilience as a Useful Concept for Climate Change Adaptation? / The Politics of Resilience for Planning: A Cautionary Note," *Planning Theory & Practice* 13, no. 2 (2012): 299–333.
6. Mick P. Kelly and Neil Adger, "Theory and Practice in Assessing Vulnerability to Climate Change and Facilitating Adaptation," *Climatic Change* 47, no. 4 (2000): 325–352; Anthony Oliver-Smith, "Anthropological Research on Hazards and Disasters," *Annual Review of Anthropology* 25, no. 1 (1996): 303–328; Dean Hardy, Richard A. Milligan, and Nik Heynen, "Racial Coastal Formation: The Environmental Injustice of Colorblind Adaptation Planning for Sea-Level Rise," *Geoforum* 87 (2017): 62–72; Natalie Osborne, "Intersectionality and Kyriarchy: A Framework for Approaching Power and Social Justice in Planning and Climate Change Adaptation," *Planning Theory* 14, no. 2 (2015): 130–151.
7. Elise Remling, "Depoliticizing Adaptation: A Critical Analysis of EU Climate Adaptation Policy," *Environmental Politics* 27, no. 3 (2018): 477–497.
8. Courtney Work et al., "Maladaptation and Development as Usual? Investigating Climate Change Mitigation and Adaptation Projects in Cambodia," *Climate Policy* 19, Suppl 1 (2019): S47–S62.

9. Robert Fletcher, "Capitalizing on Chaos: Climate Change and Disaster Capitalism," *Ephemera: Theory & Politics in Organization* 12, no. 1–2 (2012): 97–112; Glenn Fieldman, "Neoliberalism, the Production of Vulnerability and the Hobbled State: Systemic Barriers to Climate Adaptation," *Climate and Development* 3, no. 2 (2011): 159–174.
10. Deborah Harford, Nancy Olewiler, and John Richards, *Climate Change Adaptation: Linkages with Social Policy* (Vancouver, Canada: Simon Fraser University, 2010).
11. Mark Pelling, Karen O'Brien, and David Matyas, "Adaptation and Transformation," *Climatic Change* 133, no. 1 (2015): 113–127; Karen O'Brien, "Global Environmental Change II: From Adaptation to Deliberate Transformation," *Progress in Human Geography* 36, no. 5 (2012): 667–676.
12. Masroora Haque and Saleemul Huq, "Bangladesh and the Global Climate Debate." *Current History* 114, no. 771 (2015): 144–148; Shahjahan Mondal et al., "Agricultural Adaptation Practices to Climate Change Impacts in Coastal Bangladesh," in *Confronting Climate Change in Bangladesh: Policy Strategies for Adaptation and Resilience*, ed. Saleemul Huq et al. (Cham, Switzerland: Springer International, 2019), 7–21.
13. Silja Klepp, "Framing Climate Change Adaptation from a Pacific Island Perspective—The Anthropology of Emerging Legal Orders," *Sociologus* 68, no. 2 (2018): 149–170; Silja Klepp and Johannes Herbeck, "The Politics of Environmental Migration and Climate Justice in the Pacific Region," *Journal of Human Rights and the Environment* 7, no. 1 (2016): 54–73.
14. Daniel Morchain, "Why Must Climate Change Be De-naturalised and Re-politicised and What Does That Mean?" (2016), http://www.climateprep.org/stories/2016/10/13/why-must-climate-change-be-de-naturalised-and-re-politicised-and-what-does-that-mean?rq=daniel%20morchain.
15. Kyle Powys Whyte, "Is It Colonial Déjà Vu? Indigenous Peoples and Climate Injustice," in *Humanities for the Environment: Integrating Knowledge, Forging New Constellations of Practice*, ed. Joni Adamson and Michael Davis (Oxford: Earthscan, 2016), 88–104.
16. Blair Feltmate et al., *Measuring Progress on Adaptation and Climate Resilience: Recommendations to the Government of Canada* (Environment

and Climate Change Canada, 2018), https://www.csla-aapc.ca/sites/csla-aapc.ca/files/EPCARR_report.pdf.

17. Erik Swyngedouw, "Modernity and Hybridity: Nature, Regeneracionismo, and the Production of the Spanish Waterscape, 1890–1930," *Annals of the Association of American Geographers* 89, no. 3 (1999): 443–465; Donna J. Haraway, *Staying with the Trouble: Making Kin in the Chthulucene* (Durham, NC: Duke University Press, 2016).
18. David Schlosberg, *Defining Environmental Justice: Theories, Movements, and Nature* (Oxford: Oxford University Press, 2009).
19. Edward Nichol and Deborah Harford, *Low Carbon Resilience: Transformative Climate Change Planning for Canada*, ACT White Paper, Adaptation to Climate Change Team (Vancouver, Canada: Simon Fraser University, 2016). http://act-adapt.org/wp-content/uploads/2016/11/low_carbon_resilience_13.pdf.
20. World Climate Research Programme, "Global Research and Action Agenda on Cities and Climate Change Science," 2019, https://www.wcrp-climate.org/WCRP-publications/2019/GRAA-Cities-and-Climate-Change-Science-Full.pdf.
21. Emilie S. Cameron, "Securing Indigenous Politics: A Critique of the Vulnerability and Adaptation Approach to the Human Dimensions of Climate Change in the Canadian Arctic," *Global Environmental Change* 22, no. 1 (2012): 103–114.
22. Silke Beck and Martin Mahony, "The IPCC and the Politics of Anticipation," *Nature Climate Change* 7, no. 5 (2017): 311–313.
23. EnJust is a network that raises awareness for issues of environmental justice and strengthens the democratic participation of those affected by environmental problems; see https://enjust.net/uber-enjust/.

CONCLUSIONS

Ethics in Decisions on Tough Problems

GONZALO LIZARRALDE, LISA BORNSTEIN,
AND TAPAN DHAR

ARE GOOD IDEAS GOOD ENOUGH?

In 2017, the Global Facility for Disaster Reduction and Recovery (GFDRR) celebrated what it called "a decade of progress in disaster risk management."[1] GFDRR is an international partnership, established in 2006, that aims to support developing countries in disaster risk reduction and climate change adaptation. Its 2017 report made several claims: "The way the world approaches disaster risk management has evolved and matured, becoming more sophisticated and better integrated with other policies." Advances in computing, social media, and drone technology have "improved the ability of countries and organizations to predict and reduce risk." Today, "there is a greater coordination of efforts between affected countries and the international community" and by "using social media and crowdsourcing, communities are increasingly able to prepare themselves for disasters, while a range of financial instruments have been developed to reduce the impact and help them build back better." The GFDRR concluded, based on these and other observations, that "international actors are now proactively managing and incorporating disaster risk" in the adoption of the

Hyogo Framework for Action (2005), the Sendai Framework for Disaster Risk Reduction (2015), the Millennium Development Goals (2015), and the Sustainable Development Goals (2015).

The overstated optimism of the GFDRR report is perhaps surprising. Yet it illustrates a common approach adopted by international organizations, agencies, and politicians to explain problems of risk, climate change impacts, and disasters today: what these problems need is a "good idea." According to this perspective, advances in technology reduce risks, the creation of new financial instruments solves reconstruction issues, crowdsourcing helps people build back better, new international policies produce change, and global goals create consensus. Yet the GFDRR and similar organizations confuse the adoption of a good idea with its effective capacity—they presume that a first step in the right direction will produce positive change and avoid negative secondary effects.

In recent decades, some progress has been made in vulnerability reduction: infant mortality has declined; more people benefit from sanitation and health services; life expectancy has increased almost worldwide; crime and violence have decreased in several countries. But the panelists and participants in our debates have consistently argued that a more nuanced view of progress is required. As we have seen in the debates presented in this book, several problems persist across the Global North and South. We remain incapable of reducing pollution stably, while still depleting ecosystems and overexploiting resources. Wars and violence continue to displace millions of people. Poverty, inequality, and social and environmental injustices prevail worldwide.

While these problems persist, the notions of resilience, climate adaptation, community participation, technological innovation, and sustainability have become popular. They form part of the language used by politicians, donors, NGOs, funding agencies, technocrats, and citizens. But our debate participants argue that, despite the popularity of these "good ideas," not enough people have benefited.

Defenders of these popular concepts and terms might retort that they are not problematic as to the value of their ideas, but rather because not enough institutions or people have adopted them or they are being implemented with insufficient resources.

The debates presented here, however, suggest that our path forward might be more challenging. Four factors underlie the complexity regarding the use of such terms and notions: the value of their suggested ideas cannot be taken for granted; their implementation is often a major barrier; unexpected negative effects arise from "good" ideas; and the tools we have to describe and assess reality are often partial and insufficient.

CHALLENGING VALUE, IMPLEMENTATION, IMPACTS, AND DESCRIPTIONS

The value of the concepts discussed in these debates provides a first layer of complexity. Panelists and participants argued that the usefulness of these concepts can and should be challenged in theoretical, empirical, and practical ways.

Take the case of resilience. The value of resilience can be tested in theoretical terms by asking, for example, whether "rebounding" is a natural feature of human systems: Does human-made infrastructure recover autonomously in the way ecological systems do? Researchers can also challenge resilience in empirical ways: Do people in a certain community understand resilience similarly? Does resilience mean the same thing in different cultural and linguistic contexts? Finally, resilience can be challenged in practical terms: Are the precepts of resilience effectively communicated to people in a certain area? Are sufficient resources allocated to resilience programs? Our debate participants have posed similar questions regarding the notion of the fragility of cities, adaptation, community participation, emergency shelter, green certifications, and construction standards, among other subjects.

The second factor of complexity relates to the process of moving from "paper" to "practice." Our debates suggest that barriers to positive change often lie in the processes of implementation. For instance, universal construction codes and norms are perhaps noble objectives, but can they be effectively enforced in the informal sector? The idea of sheltering disaster victims in temporary units also seems positive at first, but can organizations and builders produce appropriate temporary living environments in short periods? The abstract value of codes, standards, and temporary shelter is not as fraught with difficulty as implementing them efficiently on the ground.

Unfortunately, several international organizations and consultants are currently more interested in drafting recommendations, policies, guidelines, pathways, and reports than in pursuing implementation. Implementation, particularly when influencing spatial and urban systems, often entails some or all of the following: abundant resources; personal effort, engagement, and attention to detail; conflict resolution and team-building skills; difficult negotiations with stakeholders and with funders; managerial and clerical capacity; clerical and bureaucratic work; political support; and lots of time. Creating a "good idea" and approving it in a boardroom is the easy part of the job—making it happen "on the ground" is a different story.

The secondary effects and unintended consequences of such "good ideas" are the third level of complexity. Take the case of rapid reconstruction. Building temporary solutions in urban peripheries often seems like a good idea. Land is more readily available there than in city centers, construction work is easier to put in place, and fewer conflicts with stakeholders emerge than in densely populated areas. Building on empty land avoids intervening in city centers where high land costs, limited land supply, heritage, political pressure, existing infrastructure, and urban codes are common constraints. But several studies have shown, and our debates have confirmed, that reconstruction efforts in urban peripheries lead to several negative effects. Relocated families lose the social networks they had in their original urban locations. Jobs and commercial activities are often scarce. Those who move to the outskirts have less access to public services and must spend more time commuting. Families often find that new developments lack the character and livelihood of central districts. Additionally, new residential developments in peripheries enlarge urban footprints, challenge the infrastructure's sustainability, and occupy land that could otherwise be used for agriculture or left untouched.

Many actions taken in the name of resilience and climate adaptation produce negative secondary effects—so much so that scholars have coined the term "maladaptation." Our debates highlight the prevalence of maladaptation and other mistakes made in risk reduction activities. They show how efforts intended to reduce risk can create and recreate vulnerabilities, exacerbate inequality, and result in environmental degradation. Discussions in these debates also showed that implementing environmental standards, housing refugees in camps, and relying on participation to guarantee project quality produce worrisome, and often overlooked, secondary consequences. Our panelists, therefore, insisted on the need to examine the positive and negative impacts of key advancing concepts related to risk and resilience in the short, medium, and long term.

Finally, we must also consider how various terms and concepts describe actual contexts and conditions. What do we mean when we say that residents in a certain place are "resilient"—that they survive in the face of disasters and hardships? Do they receive the support of the government or other service providers, or are they forced to operate independently to tackle established problems? Similar questions can be raised about other widely employed terms. When we label a building "sustainable," a system "adaptive," or a method "participatory," which aspects are revealed and which are overlooked?

The chapters in this book show how these and other labels only partially help us explain reality. Many affected communities do not want to be called resilient if the term implies that they will not need support and protection from authorities. Most professionals observe that sustainable construction increasingly means buildings where efforts were made to reduce energy consumption; they note that the label rarely refers to social and cultural aspects. Some participants claimed that climate adaptation might be a risky goal if it implies that humans can continue current or similar levels of pollution. In all cases, language emphasizes a particular feature of lived experience, while rendering other conditions more difficult to understand or achieve.

In chapter 2, University of Auckland professor J. C. Gaillard argues that concepts and Latin etymologies often lack the ability to fully capture the reality of people's everyday lives in dealing with what researchers commonly call "disasters." Gaillard refers to the natural limitations of language, but also to a more worrying trend identified in these debates: the unbalanced transfer of concepts and ideas resulting from unequal power relationships.

DISASTER COLONIALISM AND OTHER UNEQUAL POWER RELATIONS

The debates in this book highlight a trend in international disaster and climate policies to transfer Western experiences and ideas, attributed by participants to the Global North, to countries and communities in the Global South—specifically, developing countries. Some panelists and participants argued that this form of "disaster colonialism" often leads decision-makers, politicians, and scholars to favor the importation of ideas and approaches while disregarding local culture, knowledge, voices, and practices.

Geopolitics, capitalism, and international funders and consultants influence the deployment of aid, responses to refugees, the adoption of resilience measures, vulnerability metrics, construction codes, the implementation of green standards, and many other aspects of risk management. Some panelists contended that influences from the Global North still produce unequal power relations among practitioners, researchers, funding agencies, and organizations. They noted the persistence of "intellectual elitism" in disaster studies and practices.

These power dynamics hinder collaboration. As Thomas Fisher, an architect based in the United States, notes in chapter 1, "collective creativity and social innovation" often remain unacknowledged. Debaters suggested that a lack of knowledge-sharing and poor recognition of contributions from the Global South often aggravate injustices, facilitate international inference, and preserve power imbalances. According to Barcelona professor Carmen Mendoza Arroyo, unfavorably guided disaster-related research and practice happen due to unsubstantiated ideas and unequal power relations (national and international), which overlook culture and local knowledge and practice.

Some participants in chapter 2 and other debates challenged traditional divisions between North and South and locals and foreigners. For them, the "local" rarely exists independently from the "global." Ekatherina Zhukova, a participant, suggested that "it might be more beneficial to discuss privilege instead of distinctions between North and South," arguing that "both Northerners and Southerners can have privileges in a sense of social capital (language, finance, networks)." She thus suggested reframing our debate question as: "Is disaster-related research and practice unfavorably guided by a researcher's or practitioner's privilege?" Another participant added that we might be adopting a form of "othering" in presenting "locals" as free from global influence and thus as nonpolitical actors. Classifying people as locals and nonlocals can evidently be difficult. For instance, should the wealthy businessman, recent rural migrant, or recently arrived refugee living in a low-income neighborhood be considered locals or foreigners? To move beyond such local-foreigner, North-South, or West-and-the-Rest binaries, Mendoza-Arroyo recommends implementing horizontal learning initiatives between community organizations and broader professional networks, grounded in local ontologies and epistemologies to address such biases.

In chapter 7, some participants explored the benefits of South-South collaborations, avoiding the traditional power imbalances between the North and South. But is South-South aid a better alternative to aid provided by the North? Participants in our debates were undecided and pointed to examples in which patterns of domination and exploitation are reproduced within and among countries in the Global South. Are local actors free of the biases often associated with foreign interference? We recognized the influence of colonialism, imperialism, and neoliberalism, as well as the persistence of power imbalances, on a global scale, but we also found many of the challenges to North-South and South-South dynamics compelling. The debates suggested that power imbalances in research and practice, much like those shaping other forms of social injustice, are rooted in differences among nations and within countries, cities, communities, and organizations. The debates point to the importance of redressing unequal power relationships at all levels: between the North and South, within regions and countries, and even within communities.

UNADDRESSED VULNERABILITIES, DEPOLITICIZED ACTIONS

Discussions on how common practices often fail to address root vulnerabilities, and thus the root causes of disasters and destruction, surfaced frequently through the debates in this book. Debate participants and panelists pointed to five elements: inattention to the creation of vulnerability; the depoliticization of disaster action; issues regarding narratives of progress; the disregard of human feelings and emotions; and deficiencies in local actors' modes of participation.

The root causes of vulnerability are often dynamic and heterogeneous. They include, among others:

- Colonial legacies and postcolonial practices and their influence on patterns of land occupation, land ownership, urban form, and border definitions, as well as the importance given to some regions and social groups over others and the influence of global systems and politics in development objectives.
- Cold War–era struggles, neoliberalism, geopolitics, international military pressure, and other power dynamics and

their influence on economic policy, trade agreements, institutions, international debt, and governance mechanisms.

- Postindependence efforts in nation-building and economic growth that, though varied over time and space, often entail heavy debts with international banks and conditions placed on national policies affecting the economy, infrastructure development, public sector employment, taxation, and state responsibilities.
- Uneven or unstable revenues and weak bargaining power in global markets.
- Political corruption, cronyism, elitism, authoritarian rule, tribalism, and clientelism, which erode democracy, weaken institutions, and erode trust in authorities and governments.
- Dictatorships, civil wars, international armed conflicts, military coups, armed insurgency by dissenting groups, crime, and other forms of violence, which often weaken certain social groups and oppress communities and individuals (often according to race, gender, religious affiliation, and other conditions).
- Patriarchal structures, discrimination based on gender, race, creed, and other social injustices that marginalize communities, social groups, and individuals.
- Social and cultural dynamics—at household, community, regional, and global levels—that determine the uneven distribution of roles, duties, recognitions, land tenure, and rights and freedoms.

Yet disasters are still too frequently considered to be singular natural events. The common consequence is that the politics of disaster risk creation are neglected. Consider depoliticization in risk reduction activities and disaster reactions. Participants in chapter 8 found that policies supporting physical adaptation to the environment have helped to depoliticize risk reduction and disaster response. In other debates, scholars and practitioners concluded that narratives of sustainability and resilience also mask the political foundation of climate action and vulnerability reduction. Some argued that "green infrastructure," for instance, often causes secondary risk effects, such as gentrification and displacement. Green certifications typically mask the political, social, and cultural conditions that must be addressed to protect the environment. Quite often, risk reduction is framed as a technical, rather than a social and political, issue. Resilience policy is often presented as politically neutral, masking its tendency to benefit some groups more than others.

In general, debaters suggest that disaster risk reduction should be considered a political, rather than a technical, issue. This stance recognizes that vulnerability is constructed over time and that vulnerability reduction and disaster responses produce winners and losers, require social and political engagement, and affect territories and societies in radical and often permanent ways. More importantly, it highlights that actions taken in the face of risk and destruction cannot be separated from discussions about distributive justice.

Economic, social, and public health indicators show that progress has been made to reduce vulnerabilities in several cities and countries. For instance, construction codes, emergency protocols, and public awareness campaigns have greatly prevented destruction caused by urban fires. But participants and panelists remained uneasy using narratives of progress to describe current dynamics in climate action and risk reduction. Some participants argued that "a narrative of progress can be mobilized to ignore urgent local problems," and "scholars and practitioners must ensure this inequity does not happen." Debaters challenged the pertinence as well as the meaning of narratives of progress, asking what type of progress is claimed and at what expense it is achieved. For some, several development indicators have been achieved at the expense of ecosystems and biodiversity. Others contended that ideas of progress are often based on Northern standards and exported to countries and communities in the Global South.

Debaters found that our inability to grasp the root causes of vulnerability is often linked to our incapacity to understand the expectations, frustrations, and emotions of others. Common risk reduction practices rarely consider the deep and intimate connections between people and places. Relocation policy often ignores residents' attachment to the land and symbolic connections to the ocean, river, forests, mountains, or landscape. It often underestimates the value of social connections linking many informal settlements and low-income communities. Preventive or reactive relocation often disregards the importance of permanence: traditions, rituals, and symbolic threads that connect residents with their historical home. By focusing on immediate threats and exposure to natural hazards, many aid initiatives fail to address underlying causes of vulnerability such as poverty, marginalization, exclusion, racism, corruption, and poor planning. They also create new threats, such as the loss of livelihoods and social connections, including the associated psychological distress that accompanies forced or induced relocation.

Finally, the inadequacy of most project methods and strategies to mobilize local actors, civil society groups, and citizens hinders our capacity to address the root causes of vulnerability. Debaters found a common pattern among participatory strategies that do not transfer real decision-making power to poor and historically marginalized individuals and social groups. Mendoza-Arroyo reminds us how urban policy and design are often gender-biased. Most urban projects are still developed and led by male practitioners and decision-makers, even in the Global

North. Women's and girls' experiences are often ignored in both pre- and post-disaster planning, replicating the patriarchal structures at the core of gender-based exclusion. Mendoza-Arroyo emphasizes moral responsibility to ensure that a gendered perspective drives all phases of disaster risk reduction and recovery.

In chapter 6, debaters noted the tendency for participation to be performative rather than substantive. Participation often focuses on leaders' claims of democratic practices rather than locally situated inclusive engagement. As such, we should question the ethical value of participation and its effective capacity to redress power inequalities between local communities and economic, academic, and political elites. At the same time, it is fundamental to find ways to ensure that those most vulnerable to risks have a say in disaster risk reduction. Lorenzo Chelleri, chair of the Urban Resilience Research Network at the International University of Catalonia, concluded that "structural vulnerabilities can only be resolved through the collaboration of citizens, civil society, cities, governments, and international stakeholders."

In general, these debates have shown that to effectively reduce vulnerabilities, efforts to deal with natural hazards and environmental problems should not be separated from struggles for social and environmental justice. Reducing risk and rebuilding after disasters are not (only) technical problems—they are also political and ethical.

THE TENSIONS BETWEEN GLOBAL STANDARDS AND LOCAL PARTICULARITIES

The current consensus is that the planet is in danger. Current risks are caused by global changes and their secondary effects that pay no heed to territorial boundaries or national frontiers.

We need international action to counter contemporary threats to existence. Climate change and its global problems require concerted measures. It is urgent and unavoidable to reach an agreement on how to move forward—we need to "think globally" to reach a planetwide form of concerted action.

Not surprisingly, an increasing number of global organizations, international agencies, and transnational consultants are pushing for global policy frameworks, standards, goals, metrics, and tools to reduce carbon emissions, protect the atmosphere and natural species, design cities and buildings, and intervene territorially. They commonly argue that we need to identify comparative indicators to be able to measure change and guarantee accountability. Implementing change, they suggest, requires policy- and decision-makers to understand, measure, and compare conditions across diverse contexts.

Our debates showed that these international guidelines, protocols, certification methods, and practices are often problematic. Debaters argued that such standards often fail to address contextual particularities and that they are poorly adapted to the real needs and expectations of people "on the ground." Take the case of green certifications. Most countries in the Global South that have not independently developed their own green certification standards rely on international standards like LEED and BREEAM. In many cases, these standards do not respond to contextual needs and are ill-adapted to differing climates, the conditions of local construction and real estate sectors, and the characteristics of local forms of energy production and building technologies. David Wachsmuth, a professor at McGill University, finds no evidence that green certification systems lead to more sustainable buildings. He contends that whereas "building greener cities is an urgent task," certifying buildings with international standards "does not appear to be the most socially just or cost-effective way to accomplish that task in developing countries today." He emphasizes a need for "equity-focused urban sustainability interventions" and environmental policy.

Should all countries create their own standards and methods? Participants such as Jared Blum, who chaired the Environmental and Energy Study Institute's board of directors, believe there is no need to "reinvent the wheel." They argue that countries in the Global South can learn from industrialized nations and adopt best practices that fit local conditions. Others argue that we should "think globally and act locally," integrating global goals while paying attention to local conditions and communities. Still other debate participants advocate for the consolidation of hybrid forms of governance capable of dealing with transnational problems and responding to the needs and expectations of local communities.

Other debate commentators, such as Lorenzo Chelleri, see international agencies, consultants, and orchestrators as providing transferable tools and methods. In this perspective, international stakeholders provide alternatives to national programs on climate action and to the frequent inaction of politicians and agencies within central governments. For some participants, municipal governments should network and share their own experiences, knowledge, and results. Some participants argue that international consultants and orchestrators help connect ideas and develop metrics and instruments that can influence local policy and practices. They also help create consensus on global urgent challenges and establish a common language to deal with them.

Standards and regulations in the temporary housing sector are also problematic. The provision of housing, particularly the supply of low-cost housing, is not a simple technical or financial issue. Housing is a political subject involving complex administrative, legal, and financial processes.[2] It concerns several industries, including construction, manufacturing, real estate, and financial sectors; the dynamics of territory and availability of land; as well as national codes, local regulations, zoning, and urban planning.[3] Housing reflects cultural and social practices and is influenced by the informal sector, local traditions, history, cultural practices, and the unique conditions of places and social groups.[4] In other words, whereas housing in general presents a contextual challenge, post-disaster housing must additionally be developed under the pressure to build quickly. How can housing solutions respond to contextual characteristics while addressing international standards and objectives?

The debates in this book conclude that, regardless of whether regulations and standards are used in post-disaster reconstruction, low-cost housing, or urban practices, they should protect the poor, the most vulnerable, and historically marginalized populations and their assets, rather than penalize them.

We finish this book with a reflection on the implications of all these ideas for policy, practice, and academic work.

RADICAL OR INCREMENTAL CHANGE: A TOUGH DECISION FOR POLITICIANS, PRACTITIONERS, AND ACADEMICS

Given all the difficulties of common structures, methods, and practices in disaster risk reduction and climate change, should those interested in improving current realities try to dismantle current systems? Or should they instead try to improve the systems from within? For example, should young professionals boycott international aid institutions and practices, or should they get involved in and gradually change their operations? Should green certifications be scrapped or improved? Should scholars discard narratives of sustainability and resilience or try to recover and adapt them, nuancing their principles and extending their capacity to explain contemporary conditions? Should agencies avoid building temporary shelters and refugee camps, or should they try to enhance them so they better respond to people's needs and expectations?

In general, participants in these debates do not agree on the degree of change required to prevent, and recover from, disasters. They agree that some form of change in policy, practice, and scholarship is needed, but not on how radical that transformation should be.

All debates grappled with the issue of "throwing the baby out with the bathwater"—potentially losing something valuable by removing something not valuable. Some participants believed that several systems and social institutions are so inherently dysfunctional that they must be completely scrapped. Others believed they could be improved, and that we should aim to do so rather than seeking idealistic and utopian solutions. As editors of this book, we believe in opportunities both for stakeholders committed to radical change and for those focused on incremental, modest improvements. We need politicians interested in radical change but also require more moderate actors interested in maintaining institutions while trying to improve them. We need thinkers who can disrupt current scholarship and researchers who can refine current concepts. Today's problems call for practitioners and activists interested in disrupting the status quo and engaging with radical innovation, but they also require professionals, technocrats, and workers who can enact incremental changes within organizations and public institutions, improving current practices and conceiving possibilities for longer-term, more fundamental transformations.

There is no doubt that patriarchal structures, racism, corruption, misogyny, and other inequitable behaviors must be eradicated. However, many institutions, ideas, and cultural practices valued by stakeholders exist in both international aid and emergency and refugee responses. Most actors believe that these institutions and practices are essential in moments of crisis and that they can and should be fixed. In such cases, policy- and decision-makers must try to make informed decisions for what should be changed and what should remain stable. Researchers and think tanks must provide knowledge and evidence to facilitate new and tough decisions. They must help identify current blind spots, real inefficiencies, and problematic variables to be replaced by more effective solutions.

Tough decisions require significant sacrifices. Reductions in consumption, for instance, might cause unemployment and economic loss. Radical reductions in fossil fuel use will affect workers and investors. Politicians must have the courage to extend the rights of some and restrict the privileges of others to secure social justice. They must explain the need for such sacrifices to the public and compensate those who are vulnerable and will be affected most strongly. Leaders will surely face pressure from those who want to maintain their privileges.

Academics must create knowledge that can be used by policy- and decision-makers to cut, for instance, the types of subsidies that facilitate pollution. Knowledge is required to advance changes that reduce carbon emissions and protect people and ecosystems. Leaders and policymakers must have both the conviction to advance radical transformations and the patience to produce incremental changes. In both cases, change must be based on science drawing on academic and local knowledge, rather than on wishful thinking, appealing buzzwords, ideology, biases, or untested hypotheses.

THE NEED FOR AN ETHICAL FRAMEWORK

Panelists and participants voiced an urgent need for alternative and more ethical ways of considering and responding to risk, climate change, and disasters. This shift in approach must consider political and socioecological specificities and dynamics, individual and community voices, secondary effects and unintended consequences, traditional knowledge, and local conditions. We need an ethical framework capable of helping us navigate the dilemmas and tough decisions associated with disruption and destruction. This ethical lens must be informed by a thorough understanding of inequalities, power imbalances, privileges, and social and environmental injustices. In other words, disaster scholarship, policies, and practices should originate in science, experiential knowledge, and vernacular narratives. We need an empathetic consideration of people living in disaster-prone contexts. It is important to connect disaster risk reduction policy and practices to local knowledge and experience, while scaling up rigorous research and experiential learning.

J. C. Gaillard recommends acknowledging how dominant epistemologies—what is considered valid knowledge and what is not—influence policy and standard practices. Silja Klepp, a professor at Kiel University, warns us that ignorance often increases the risk of inflicting epistemic violence and injustices on the most vulnerable. Both argue that scholars can and must reveal the root causes of disasters and injustices. Academics from the Global South can be involved in engaged, participatory research to help voice their perspectives, as well as the needs and aspirations of the most vulnerable.

More humility is also required from all scholars, from the Global North and South. As some debaters argued, academics must acknowledge not only what they know but also the limitations of their knowledge. Overconfidence prevents scholars from seeing blind spots and anticipating secondary effects. Such observations apply equally, if not more so, to the policymakers and practitioners involved in disaster response, recovery, and risk reduction. Their good intentions must be met with a serious commitment, one that goes beyond "do no harm," if they wish to advance meaningful and just contributions to the well-being of humans and nonhumans alike.

Disasters are socially constructed. Our preconceived ideas about risk, our perspectives on the future, our immediate and long-term interests, and our intentions create contradictions that are shaped by moral dilemmas and the limits of our cognitive capacity to understand reality. Our histories, the power relationships defining our communities, and our cultural and social institutions influence our actions and understanding of problems and potential solutions. Two fundamental directions must therefore be considered when constructing ethical frameworks for tackling disasters and risks: first, to frame and reframe present issues; and second, to challenge the value, effectiveness, pertinence, and impacts of our actions.

Facing current and future challenges will call for difficult decisions. The type of change required will demand significant sacrifices. Today more than ever before, we require an ethical framework to deal with risk, disasters, and climate change. A collective effort is required to avoid the adverse effects of colonialism, exclusion, marginalization, racism, segregation, and other social and environmental injustices. On a planet at risk, humans are increasingly threatened by natural hazards, and natural ecosystems are threatened by human activity. Our best choices moving forward are to reframe common environmental and social problems and to reexamine our ideas from the viewpoint of the most vulnerable people and species.

NOTES

1. Global Facility for Disaster Reduction and Recovery (GFDRR), "A Decade of Progress in Disaster Risk Management," 2017, https://www.gfdrr.org/sites/default/files/10%20Years%20DRM%20Development%20DRAFT.pdf, 5.
2. Gonzalo Lizarralde, *The Invisible Houses: Rethinking and Designing Low-Cost Housing in Developing Countries* (London: Routledge, 2014).
3. Gonzalo Lizarralde, "Stakeholder Participation and Incremental Housing in Subsidized Housing Projects in Colombia and South Africa," *Habitat International* 35, no. 2 (2010): 175–187; Gonzalo Lizarralde and David Root, "The Informal Construction Sector and the Inefficiency of Low-Cost Housing Markets," *Construction Management and Economics* 26, no. 2 (2008): 103–113.
4. Lisa Bornstein, "Introduction to Special Section on the Informal Sector," *Berkeley Planning Journal* 7, no. 1 (1992): 121–123.

CONTRIBUTORS

Kamel Abboud is an architect and a graduate of the Académie Libanaise des Beaux Arts, where he has taught since 1997. Between 1987 and 1994, he worked in Paris as an architect with three architectural firms. He then spent three years as an associate senior architect with Architectural Design Unit at Basile, Homsi, and Associates in Beirut. In 1997, he established his own independent architectural studio, AK-Architects, based in Beirut. In September 2016, he presented "A Century of Migrations and Urban Disturbances in Lebanon Between 1916 and 2016" at the New Cities and Migrations International Workshop at the Università degli Studi in Florence.

Brian Aldrich is an emeritus professor in the Department of Sociology and Criminal Justice at Winona State University and previously taught at the University of Minnesota. He acted as a block group organizer with the Lincoln Park Conservation Association in Chicago (1962–1963) and as the director of workers' education for the Mindanao Federation of Labor in the Philippines (1963–1965). He also worked for the Presbyterian Institute for Industrial Relations in Chicago, where he researched urban ministries (1965–1966). Aldrich has published several research papers on housing and community organization in Southeast Asian cities and is the co-editor of three books with Ranvinder S. Sandhu: *Housing in Asia: Problems and Perspectives* (1990), *Housing the Urban Poor: Policy and Practice in Developing Countries* (1995), and *Housing the Urban Poor in Developing Countries* (2015). He received his Ph.D. in sociology, social organization, and demography with a minor concentration in urban and regional studies at the University of Wisconsin-Madison in 1972.

Daniel Aldrich is the director of the Security and Resilience Studies Program and a professor of political science and public policy at Northeastern University. He has spent more than five years carrying out fieldwork in Africa, Asia, and the Middle East. His research has been funded by the National Science Foundation, the Fulbright Foundation, and the Abe Foundation. Aldrich has published five books and more than seventy-five peer-reviewed articles, and he has written op-eds for the *New York Times*, CNN, *Asahi shinbun*. He also has appeared on popular media outlets, including CNBC, MSNBC, NPR, and HuffPost.

Jared Blum has chaired the board of directors for the Environmental and Energy Study Institute since December 2010. From 1990 to 2016, Blum acted as the president and chief executive officer of the Polyisocyanurate Insulation Manufacturers Association. As vice chair of the Business Council for Sustainable Energy since 1997, Blum has participated in numerous United Nations climate negotiations, demonstrating business support for the Conferences of the Parties process.

Camillo Boano is an architect, urbanist, and educator. He is a senior lecturer in the Bartlett Development Planning Unit at University College London, where he directs the MSc in Building and Urban Design in Development program. He is also co-director of the UCL Urban Laboratory. Boano has more than twenty years of experience in research, design consultancy, and development work in South America, the Middle East, Eastern Europe, and Southeast Asia. As an academic interested in practice, he approaches critical architecture, spatial production and transformation, and urbanism with a focus on the exceptional circumstances of disasters, conflicts, and informality. He holds a master's degree in urban development and a Ph.D. in planning from Oxford Brookes University.

Lisa Bornstein is an associate professor in and director of the School of Urban Planning at McGill University. Her research and teaching explores the politics of planning, examining decision-making processes around international aid, urban management, economic development, and environmental policy. She has a particular interest in public deliberations in moments of environmental, economic, or social disruption; the policies, plans, and everyday practices that result; and their consequences for different groups, places, and systems. She has experience as a researcher, consultant, and educator in the Americas, Africa, and Europe, managing projects funded by, among others, the Canadian International Development Research Centre, Social Sciences and Humanities Research Council of Canada, and Fonds de recherche du Québec. Throughout her work, she aims to link participatory engaged research to academic knowledge and to real-world outcomes. She is a principal researcher with the Œuvre Durable network. She publishes broadly, with articles and chapters on comparative spatial policy, social dynamics, and the quality of urban environments.

Christopher Bryant is an adjunct professor in the Department of Geography at Université de Montréal. He is also adjunct professor in the School of Environmental Design and Rural Development at the University of Guelph, Ontario. Bryant holds a Ph.D. on the transformation of agriculture around Paris from the London School of Economics and Political Science. For more than the last twenty years, his research in the field of development has focused on community participation, sustainable community development, rural development, land use planning, strategic management and planning of development, and the adaptation of human activities to climate change. His research findings are presented in about 200 journal articles, more than 500 conference papers, over 160 book chapters, and 31 books.

Lorenzo Chelleri is the director of the international master's program in city resilience design and management and the chair of the Urban Resilience Research Network (URNet) at the International University of Catalonia. With a background in urban and regional planning, environmental policy, and urban geography, his research and teaching activities address management and planning processes related to city resilience governance. Interested in the complex relationship between urban resilience and sustainability, he is particularly concerned with the application of resilience in cities and trade-offs among social, environmental, spatial, and temporal aspects of urban resilience. Lorenzo collaborates with the European Environment Agency and has conducted research in Mexico, Bolivia, Morocco, Europe, and Asia.

Jeff Crisp is an associate fellow in international law at Chatham House and has held senior positions with the United Nations High Commissioner for Refugees (head of policy development and evaluation), Refugees International (senior director for policy and advocacy), and the Global Commission on International

Migration (director of policy and research). He has also worked for the Independent Commission on International Humanitarian Issues, the British Refugee Council, and Coventry University. Crisp has firsthand experience in humanitarian operations around the world and has published and lectured widely on refugee and migration issues. He holds a master's degree and Ph.D. in African studies from the University of Birmingham.

Tapan Dhar is an assistant professor in the School of the Environment at Trent University, specializing in the human dimensions of climate change adaptation, environmental justice, environmental planning, urban design, sustainable urban settlements, urban resilience and participatory planning. He has taught at several institutions, including the University of Oregon, McGill University, Concordia University, and Khulna University. Dhar received his Ph.D. at the University of Waterloo. He has authored numerous journal articles and book chapters, including a contribution to the Intergovernmental Panel on Climate Change (IPCC) Working Group II's Sixth Assessment Report.

Thomas Fisher is a professor in the School of Architecture, the director of the Minnesota Design Center, and the former dean of the College of Design at the University of Minnesota. Voted four times a top-25 design educator in the United States by DesignIntelligence, he is one of the most published authors in his field, having written 12 books, over 75 book chapters, and over 470 articles in major publications, including *Designing to Avoid Disaster: The Nature of Fracture-Critical Design* (2013) and *Designing Our Way to a Better World* (2016). His most recent book, *Space, Structures, and Design in a Post-Pandemic World* (2022), addresses the impact that pandemics have had on communities historically and what impacts we might expect in the wake of the COVID-19 pandemic.

J. C. Gaillard is *ahorangi*/professor of geography at Waipapa Taumata Rau/University of Auckland in Aotearoa, New Zealand. He is also a member of the board of directors of the Center for Disaster Preparedness in the Philippines. His work focuses on power and inclusion in disaster studies and develops participatory tools for engaging minority groups in disaster risk reduction, with an emphasis on ethnic and gender minorities, prisoners, children, and homeless people.

Deborah Harford is senior adviser and former executive director of the Adaptation to Climate Change Team (ACT) at Simon Fraser University. At ACT, she helped develop its initiative and partnerships with the public, private, and philanthropic sectors, as well as managed the program. A leader on integrating adaptation and mitigation through low carbon resilience, she directed and produced ACT's research and policy recommendations to advance sustainable climate action strategies at all levels and sectors of government, as well as communication and promotion of the program's outcomes. Harford is a contributing author on the Cities and Towns chapter of the 2022 Canadian national climate risk assessment and has participated in numerous national initiatives, including iEnvironment and Climate Change Canada's Expert Panel on Adaptation and Resilience Results, the Council of Canadian Academies' Expert Panel on Canada's Top Climate Change Risks, the Technical Working Group of the Canadian Centre for Climate Services, the Infrastructure and Buildings Working Group of Canada's National Adaptation Platform, and the Expert Adaptation Panel of the new Canadian Institute for Climate Choices.

Craig Johnson is professor of political science at the University of Guelph, Ontario, where he teaches environmental politics, sustainable development, humanitarian policy, and global environmental regimes. His research lies in the field of global environmental governance and focuses primarily on the role of cities and transnational city networks in reducing the global carbon footprint. Craig is the author of *The Power of Cities in Global Climate Politics* (2018) and *Arresting Development: The Power of Knowledge for Social Change* (2009). He is also a senior fellow with the Global Cities Institute at the University of Toronto and has taught at the London School of Economics, the School of Oriental and African Studies, University College London, and the University of Oxford.

Jonathan Joseph is a professor of politics and international relations in the School of Sociology, Politics, and International Studies at the University of Bristol. His research applies concepts of resilience and governmentality to a range of fields, such as security strategy, state-building, and EU politics. He is particularly interested in how governmentality works in different contexts and whether it can explain international or global forms of governance. His current projects involve a study of resilience in EU and international policymaking. His books include *Varieties of Resilience: Studies in Governmentality* (2018) and *Wellbeing, Resilience, and Sustainability* (with J. Allister McGregor, 2019).

Ilan Kelman is a professor of disasters and health at University College London and a professor at the University of Agder, Norway. His research focuses on intersections between disasters and health, including the integration of climate change into disaster research and health research. It covers disaster diplomacy and health diplomacy, island sustainability involving safe and healthy communities in isolated locations, and risk education for health and disasters.

Silja Klepp is professor of human geography at Kiel University, where she holds the UNESCO co-chair for integrated marine science. Her research group examines human–environment relations in the Anthropocene. Her current research on climate change adaptation, coastal erosion, and disaster governance integrates postcolonial perspectives and critical theories to explore how social and cultural diversity can be involved in adaptation and disaster governance and ensure the self-determination of impacted communities. She has worked in the field in countries such as Cabo Verde, Kiribati, Vanuatu, New Zealand, Italy, Libya, Malta, and Zambia. Together with colleagues, Silja Klepp founded the transdisciplinary environmental justice network EnJust.

Anna Konotchick is the director of housing and human settlements at the Asia Pacific Regional Office of Habitat for Humanity International in Manila. She leads her team in the following fields: construction, engineering, housing, urban planning, land tenure, disaster response, recovery, and resilience. From 2014 to 2018, she was the Canaan Program manager for the American Red Cross in Haiti. She oversaw directly implemented and partner-implemented projects in the Canaan urban development and resilience portfolio. Anna previously worked for the World Bank, providing technical assistance on beneficiary satisfaction in Haitian housing reconstruction. She received her bachelor's in science from MIT, and two master's degrees in architecture and city planning from the University of California, Berkeley.

Gonzalo Lizarralde is a professor in the School of Architecture at Université de Montréal. He also holds the Université de Montréal's Fayolle-Magil Construction Chair in Architecture, Built Environment, and Sustainability and is the director of the IF Research Group (GRIF) and the Canadian Disaster Resilience and Sustainable Reconstruction Research Alliance (Œuvre Durable). His work focuses on understanding risk, low-cost housing, and informality in urban settings. In addition to co-founding i-Rec, an international network of specialists in disaster risk reduction and reconstruction, he is the author or editor of many publications on these topics, including *Unnatural Disasters: Why Most Responses to Risk and Climate Change Fail But Some Succeed*, *The Invisible Houses: Rethinking and Designing Low-Cost Housing in Developing Countries*, and *Rebuilding After Disasters: From Emergency to Sustainability.* Lizarralde is a member of the College of New Scholars, Artists, and Scientists of the Royal Society of Canada.

Michael Mehaffy is a researcher, educator, urban and building designer, architectural theorist, and urban philosopher. His work focuses on the dynamics of urban growth, urban networks, compact walkable cities, and new tools to exploit their social, ecological, and economic advantages. He holds a Ph.D. in architecture from Delft University of Technology and has held research and teaching appointments at seven graduate institutions in six countries. Mehaffy sits on the editorial boards of two international urban design journals and has published or contributed to over twenty books. His writing and interviews frequently appear in professional and trade publications, as well as in *The Guardian*, *Scientific American*, *Voice of America*, *The Atlantic*, among others. He is currently a senior researcher at the KTH Royal

Institute of Technology in Stockholm and the managing director of Sustasis Foundation, a urban think tank based in Portland, Oregon.

Carmen Mendoza-Arroyo is an architect and a Ph.D. student in urban design and planning. She is an associate professor and assistant director at the School of Architecture at Universitat Internacional de Catalunya (UIC Barcelona), where she is also the director of the Master's in International Cooperation program: Sustainable Emergency Architecture. Her recent research focuses on urban reconstruction and resilience in the field of emergency architecture as well as urban integration strategies for displaced populations and refugees in Catalonia, Colombia, Ecuador, Brazil, Greece, India, Lebanon, and Perú. Mendoza-Arroya has published her research extensively in articles and books and has edited several books.

Graham Saunders (1961–2017) made significant contributions to the design, management, and technical support of shelter and settlement relief and development programs. A UK-trained architect with a background in project design and management, he gained extensive experience in shelter programming in Africa, South and Southeast Asia, Eastern Europe, and Central Asia, the Middle East, Latin America, and the Caribbean. Saunders served as the head of shelter and settlement for the International Federation of Red Cross and Red Crescent Societies in Geneva until his death in 2017 and established the Global Shelter Program. His work left a profound and lasting impact on global disaster response efforts.

Jason von Meding is an associate professor at the University of Florida and a founding faculty member of the Florida Institute for Built Environment Resilience (FIBER). Before moving to the United States, he established the Disaster and Development Research Group at the University of Newcastle, Australia. He obtained his Ph.D. from Queen's University of Belfast, where he worked for three years as a faculty member. His research considers the social, political, economic, and environmental injustices that cause people across global societies to be marginalized and exposed to greater disasters. Von Meding is the writer and an executive producer of an upcoming DEVIATE documentary and often contributes to the online publication *The Conversation*.

David Wachsmuth is the Canada Research Chair in Urban Governance at McGill University, where he is also an associate professor in the School of Urban Planning. He directs UPGo, the Urban Politics and Governance group at McGill, a team of researchers investigating pressing urban governance problems related to economic development, environmental sustainability, and housing markets. He is the co-lead of the Adapting Urban Environments for the Future theme of the McGill Sustainability Systems Initiative. Wachsmuth has published widely in urban studies, planning, and geography, and his work has been covered extensively in national and international media, including the *New York Times*, the *Wall Street Journal*, the Associated Press, and the *Washington Post*.

Edmundo Werna has worked for more than thirty-five years on different aspects of urban development, with particular attention to municipal management, livelihoods, and housing. He joined the United Nations' International Labour Office (ILO) in 2004. One of his current responsibilities is to provide technical support to the Habitat III process on behalf of the ILO. Prior to joining the ILO, he designed and implemented the urban development agenda of the United Nations Development Programme's UN Volunteers Programme. He also undertook consultancies for several organizations, including various local governments, the World Health Organization, the European Commission, the World Bank, UN-Habitat, and UNCDF/UNDP. He has held research and lecturer positions in research and lecturing in British, Brazilian, and Italian universities, as well as at the Woodrow Wilson International Center for Scholars in Washington, DC.

INDEX

"*Thinking in Transit* speaks to the heart of our transitory existence. The authors bri us on a wondrous journey, showing that some of our best thoughts are mobile, itin ant, searching—odysseys of both mind and body. This is a brilliant and timely book.

—RICHARD KEARNEY,
CHARLES SEELIG PROFESSOR IN PHILOSOPHY, BOSTON COLLEGE

"Craig and Casey share a singularly original project, interweaving personal recolle tions with insightful philosophical ruminations. They challenge the assumption th thinking must be solitary and sedentary, bringing thought back into the motion a emotion of everyday life."

—DAVID MICHAEL KLEINBERG-LEVIN, PROFESSOR EMERITUS,
DEPARTMENT OF PHILOSOPHY, NORTHWESTERN UNIVERSITY

"From their form to their moving details, these essays slide between authors, acro elemental fields, and in and out of vessels, from boats to swings to sleds. And alo the way, we find ourselves invigorated by a transformed feel for where we've alwa been: on the move."

—JOHN LYSAKER, WILLIAM R. KENAN UNIVERSITY PROFESSOR AND
DIRECTOR OF THE CENTER FOR ETHICS, EMORY UNIVERSITY

"A deeply meditative book. Craig's and Casey's voices—sometimes blended, som times separate—eloquently evoke the wonder and significance of everyday mov ments: swimming, falling, skating, and flying, just to name a few. I'll never think taking the ferry the same way again!"

—SHANNON SULLIVAN, PROFESSOR OF PHILOSOPHY AND HEALTH
PSYCHOLOGY, UNIVERSITY OF NORTH CAROLINA AT CHARLOTTE

"Captivating reflections from two thinkers and visual artists on the nature of water, a earth, fire—the elements of existence. As the authors travel via trains, planes, and ca or even just inhabit gardens beyond the cloistered walls of academia, they develop singular style of painterly thinking."

—CYNTHIA WILLETT, SAMUEL CANDLER DOBBS PROFESSOR
OF PHILOSOPHY, EMORY UNIVERSITY

MEGAN CRAIG is associate professor of philosophy at the State University of Ne York at Stony Brook as well as an artist and essayist.

EDWARD S. CASEY is distinguished professor emeritus of philosophy at the Sta University of New York at Stony Brook and past president of the American Philosoph cal Association.

Cover design: Milenda Nan Ok Lee   Cover photo: Megan Craig

Columbia University Press | New York
CUP.COLUMBIA.EDU

GPSR Authorized Representative: Easy Access System Europe, Mustamäe tee 50, 10621 Tallinn, Estonia, gpsr.requests@easproject.com

www.ingramcontent.com/pod-product-compliance
Lightning Source LLC
Jackson TN
JSHW020424140825
89344JS00008B/271